Industry 4.0 Compact—How Technologies Are Transforming Our Economy and Our Companies

Walter Huber

Industry 4.0 Compact—How Technologies Are Transforming Our Economy and Our Companies

Transformation and Change of the Entire Organization

Springer

Walter Huber
Haar, Germany

ISBN 978-3-658-51028-2 ISBN 978-3-658-51029-9 (eBook)
https://doi.org/10.1007/978-3-658-51029-9

This book is a translation of the original German edition "Industrie 4.0 kompakt – Wie Technologien unsere Wirtschaft und unsere Unternehmen verändern" by Walter Huber, published by Springer Fachmedien Wiesbaden GmbH in 2018. The translation was done with the help of an artificial intelligence machine translation tool. A subsequent human revision was done primarily in terms of content, so that the book will read stylistically differently from a conventional translation. Springer Nature works continuously to further the development of tools for the production of books and on the related technologies to support the authors.

This Springer imprint is published by the registered company Springer Fachmedien Wiesbaden GmbH, part of Springer Nature.
The registered company address is: Abraham-Lincoln-Str. 46, 65189 Wiesbaden, Germany

Preface to the Second Edition

Our companies and the entire economy are currently undergoing a massive process of transformation and change. Covid-19 has greatly accelerated these changes and ruthlessly exposed weaknesses in existing structures. Further crises have intensified this trend. Overall, this process of change is leading to completely new demands on companies and the entire economy. The pressure to innovate and its dynamics have increased significantly, accompanied by changing ecosystems and, consequently, value chains as well as production and work processes. Novel products, services, and thus also new business models are both the drivers and the result of this transformation. The pandemic has forced us to adopt new ways of working and significantly altered leadership and decision-making mechanisms and processes. During the pandemic, working from home became the "new normal." Traditional leadership styles are reaching their limits under these changed conditions and must be rethought.

In summary, we must come to terms with the fact that uncertainties and unknowns will shape both our professional and private lives in the near future. As a result, experiences from the past lose their validity and applicability—at least in part. For companies, this creates the necessity for new decision-making and thinking patterns, as well as a significantly altered shift in perspective. Specifically, companies must learn to operate successfully in a dynamic and highly volatile environment and thus manage risks effectively.

Digital technologies can help us successfully manage this process of change. At the same time, however, they are also drivers of change. Their usefulness was demonstrated during the pandemic (for example, in the form of communication solutions). Without them, continuing to work from home would have been impossible.

These technologies also demonstrate their Janus-faced character in that they are both drivers and vulnerabilities at the same time. Cyber-attacks have surged in parallel with the use of these technologies. This applies not only to private and enterprise environments, but increasingly also in the field of production.

The rapid pace of change is also evident in the fact that, since the publication of the first edition at the end of 2017, new technologies have emerged. These include, among others, 5G and blockchain. The topic of cyber security has already been addressed. Furthermore, existing technologies, such as AI, have matured significantly.

Furthermore, the development of an active learning architecture is increasingly becoming a critical success factor. Only if the establishment of a learning architecture is successfully combined with active change management can smart technologies generate corresponding economic added value. Ultimately, the goal is to foster a digital mindset within companies. Simply introducing new technologies does not in itself create added value—they must, quite simply, be used—and used successfully.

Munich
in Autumn 2023

Dr. Walter Huber

Preface to the First Edition

The introduction of the first ideas surrounding Industry 4.0 is only six years old, yet already profound changes are visible in the German industrial and corporate landscape. This demonstrates the dynamism and explosive force of the current transformations taking place. These changes are multifaceted—just like the German industrial landscape itself. At least for now, there are no universal solutions for this transformation. Experimentation and a willingness to accept gaps in planning are more in demand here than meticulous preparation. The changes are advancing at a rapid pace, and anyone who does not want to be left behind by digitalization would be well advised to embark on the path of transformation as quickly as possible. Speed is of the essence.

The changes affect all industries and companies of all sizes—even down to the one-person wine merchant who now conducts most of his business through online sales to customers worldwide. These changes are therefore not limited to production, but rather encompass all areas of the company—from product development and the design of new smart products, to production with a smart factory, and on to new services and innovative business models. The company's own network is under just as much pressure as the companies themselves. "Nothing is as constant as change" is becoming increasingly true within companies.

Individual companies and the German economy as a whole must develop entirely new concepts and competencies in order to continue to exist even five years from now. Digital processes follow a completely different dynamic and exhibit a significantly different anatomy compared to proven and established approaches. On the other hand, the successes achieved through the many Lean transformations must not be lost. Until now, they have been a guarantee for cost-effective production in the high-wage country of Germany. The next stage of development—successfully initiating and implementing digital transformation—must now be achieved. It is essential to transform Germany's innovative strength and creativity into the digital age (which has long since begun). In the future, a sole focus on technology will no longer suffice; new business models must be developed and successfully positioned in the market.

Germany is currently utilizing only a fraction of its digital potential. This must change. Above all, the German Mittelstand, as the backbone of the entire economy, must live up to its responsibility and at least "shift up a gear" in its efforts to manage the ongoing transformation process. "Old habits" must be abandoned and new approaches embraced. Germany's competitiveness—and indeed our future as an economic nation—depend on the Mittelstand.

This book does not claim to provide a comprehensive account of the topics of digitalization and Industry 4.0. Such an undertaking would require a multi-volume work—and would be pointless at the same time. Technologies and their significance change far too quickly. Moreover, there is no universal recipe for digital transformation—just as in many areas of life. Rather, the book aims to "shed light on the darkness" and spark curiosity about these topics. With combined efforts within the company, digital transformation can succeed. The important thing is to get started. This book aims to contribute to that.

The first part of the book deals with the fundamentals. Here, all relevant terms related to digitalization and Industry 4.0 are explained, ranging from disruptive to evolutionary technologies. The second part gets down to the nitty-gritty—that is, it describes the necessary changes within companies. This begins with product development and new products, continues through production, and concludes with new business models. Organizational changes within companies must not be overlooked. This is followed by a presentation of countermeasures.

The aim of this book is to provide a concise yet comprehensible introduction to the topic of digitalization and Industry 4.0 from a management perspective. It is intended to encourage readers to engage with the subject. There are significant opportunities, but they must also be seized. In particular, the topic of business models is too often neglected in our highly technology-driven country.

Numerous examples from a wide range of industries and company sizes illustrate the individual topics and also provide an economic overview of the existing potential.

Of course, one can also try to avoid or "sit out" the topic of digitalization and Industry 4.0—at least in the short term. In the long run, however, this approach is a dead end. The path into the "unknown" is far less daunting than is often assumed. The goal is to discover new things without carelessly discarding proven methods. Many of the seemingly new ideas have, in their fundamentals, been known for some time. The combination of individual technologies, especially when linked with new business models, creates an explosive mix. The task is to develop your own "magic potion" in order to be able to compete with American and, increasingly, Chinese companies in the future. So, let's get to work and take action!

Munich
in November 2017

Dr. Walter Huber

Contents

1 **Motivation for Another Industry 4.0 and Digitalization Book** . . . 1
2 **Drivers of Change** . . . 3
References . . . 12
3 **What Is It About? Definition and Classification** . . . 13
References . . . 17
4 **Technologies That Revolutionize** . . . 19
4.1 Data, Software, and Smart Data . . . 20
4.1.1 Technical Description . . . 20
4.1.2 Consequences for Management . . . 21
4.1.3 Economic Relevance . . . 23
4.2 Big Data and Data Lake . . . 24
4.2.1 Technical Description . . . 25
4.2.2 Consequences for Management . . . 27
4.2.3 Economic Relevance . . . 28
4.2.4 Risks . . . 30
4.3 Digital Twin . . . 30
4.3.1 Economic Relevance . . . 32
4.3.2 Risks . . . 32
4.4 Artificial Intelligence/Machine Learning . . . 33
4.4.1 Technical Description . . . 34
4.4.2 Consequences for Management . . . 36
4.4.3 Economic Relevance . . . 36
4.4.4 Risks . . . 37
4.5 Cyber Physical Systems . . . 38
4.5.1 Technical Description . . . 39
4.5.2 Consequences for Management . . . 39
4.5.3 Economic Relevance . . . 40
4.5.4 Risks . . . 40

4.6 Sensitive Robots 41
4.6.1 Technical Description 41
4.6.2 Consequences for Management 45
4.6.3 Economic Relevance 46
4.6.4 Risks 48
4.7 Additive Manufacturing 49
4.7.1 Technical Description 50
4.7.2 Consequences for Management 50
4.7.3 Economic Relevance 52
4.7.4 Risks 52
4.8 Platforms 53
4.8.1 Technical Description 54
4.8.2 Consequences for Management 56
4.8.3 Economic Relevance 56
4.8.4 Risks 57
References 58
5 Technologies That Evolve 61
5.1 Cyber Security 61
5.1.1 Technical Description 62
5.1.2 Consequences for Management 64
5.1.3 Economic Relevance 64
5.2 Vertical and Horizontal Integration 65
5.2.1 Technical Description 65
5.2.2 Consequences for Management 67
5.2.3 Economic Relevance 67
5.3 Blockchain 68
5.3.1 Technical Description 68
5.3.2 Consequences for Management 69
5.3.3 Economic Relevance 69
5.4 Autonomous Transport Units 70
5.4.1 Technical Description 70
5.4.2 Consequences for Management 71
5.4.3 Economic Relevance 72
5.5 Real Time Enterprise and Mobile Solutions 72
5.5.1 Technical Description 72
5.5.2 Consequences for Management 73
5.5.3 Economic Relevance 73
5.6 Digital Factory 74
5.6.1 Technical Description 74
5.6.2 Consequences for Management 75
5.6.3 Economic Relevance 76

5.7 Assistance Systems . . . 76
5.7.1 Technical Description . . . 77
5.7.2 Consequences for Management . . . 79
5.7.3 Economic Relevance . . . 80
5.8 Cloud . . . 80
5.8.1 Technical Description . . . 81
5.8.2 Consequences for Management . . . 83
5.8.3 Economic Relevance . . . 83
5.9 5G/6G . . . 84
5.9.1 Technical Description . . . 84
5.9.2 Consequences for Management . . . 85
5.9.3 Economic Relevance . . . 85
5.10 Metaverse . . . 85
5.10.1 Technical Description . . . 86
5.10.2 Consequences for Management . . . 86
5.10.3 Economic Relevance . . . 86
5.11 Quantum Computing . . . 87
5.11.1 Technical Description . . . 87
5.11.2 Consequences for Management . . . 88
5.11.3 Economic Relevance . . . 88
References . . . 88
6 New Products Change the World and Companies . . . 91
6.1 Transformed Products . . . 92
6.1.1 Extended Functionality . . . 93
6.1.2 Networking and Communication . . . 94
6.1.3 Data . . . 96
6.1.4 Digital Services and Platforms . . . 97
6.1.5 Human-Machine Interaction . . . 99
6.1.6 Transformation: Development and Definition of New Functionalities . . . 100
6.2 Novel Product Development Process . . . 101
6.3 Changed Competitive Situation . . . 102
References . . . 104
7 The Smart Way of Production . . . 105
7.1 Smart Factory . . . 106
7.1.1 The Part Controls the Process . . . 107
7.1.2 Robotics . . . 110
7.1.3 Digital Twin . . . 110
7.1.4 Full Transparency through Logistics 4.0 . . . 113
7.1.5 Digital Supply Chain . . . 115
7.2 Integration and IT . . . 116

7.3 The New Role of Humans in Production . . . 117
References . . . 118
8 Opportunities Through New Business Models . . . 119
8.1 Backgrounds for Change . . . 122
8.2 Characteristics of Digital Business Models . . . 124
8.3 Impacts on Existing Business Models . . . 126
8.4 Evolutionary: Further Development of Existing Business Models . . . 127
8.5 Disruptive: Development of New Digital Business Models . . . 128
8.5.1 Personalization and Individualization . . . 129
8.5.2 CPS . . . 130
8.5.3 Service Versus Product . . . 130
8.5.4 Success-Based Approaches . . . 131
8.5.5 Selling and Using Data . . . 132
8.5.6 Platforms . . . 132
8.5.7 Digital Ecosystems . . . 134
8.6 Transformation of Business Models . . . 135
8.7 Challenges of Digital Business Models . . . 137
8.7.1 Assessment of Technologies . . . 137
8.7.2 Changed Customer Interaction . . . 138
8.7.3 Building Digital Trust . . . 138
8.7.4 Handling Data and Data Quality . . . 139
8.7.5 Infrastructure . . . 140
8.7.6 Dealing with Legacy Systems . . . 140
8.7.7 Organizational Aspects . . . 141
8.7.8 Political Aspects . . . 142
References . . . 142
9 New Companies and Leaders . . . 145
9.1 Company Transformation . . . 146
9.1.1 Organizational Aspects . . . 148
9.1.2 Technological Aspects . . . 150
9.1.3 The New Role of the HR Department . . . 152
9.1.4 Transformation Process . . . 153
9.2 Transformation of Leadership Behavior and Managers . . . 156
9.2.1 Characteristics of Digital Leaders . . . 156
9.2.2 Leadership Behavior . . . 157
9.2.3 Decision-Making . . . 158
9.3 Employee Transformation . . . 159
9.3.1 Indirect Areas . . . 160
9.3.2 Direct Areas . . . 161
9.3.3 Approaches for Action . . . 162
References . . . 163

10 Risks From New Competition 165
10.1 Examples of New Competition 166
10.2 Causes of New Competitors 167
10.3 Identification of New Competitors 169
10.4 New National Competitors 171
10.4.1 China 171
10.4.2 France 172
10.4.3 USA 173
10.4.4 Switzerland 173
10.5 Defensive Measures 174
10.5.1 Self-Cannibalization 174
10.5.2 Building Your Own Digital Ecosystem 175
10.5.3 Leveraging Your Own Strengths 176
10.5.4 Organizational Measures 176
References 177
11 Quo Vadis Germany 179
11.1 Critical Review 180
11.2 What Needs to Be Done? 182
11.3 What Happens Next 184
References 185

Motivation for Another Industry 4.0 and Digitalization Book

1

Industry 4.0 has now become part of operational practice. The number of successful implementations, for example in the form of new products and innovative production approaches, is increasing rapidly every day. All relevant industrialized nations have launched and, in some cases, already implemented corresponding programs. The naming conventions differ from country to country. In Germany, a variety of funding programs have been and are being established to support the digital transformation. These programs focus on different priorities, such as technological aspects (e.g., the use of artificial intelligence) or industry-specific topics (e.g., the automotive industry).

This approach is being driven by the imperative for sustainability. In addition, Covid-19 has dramatically changed our working world and, consequently, the way we collaborate. Altogether, this has created a conglomerate of external influencing factors that act as accelerators for Industry 4.0.

The projected economic potential is enormous and has been confirmed in recent years by practical examples. Therefore, an in-depth engagement with the topic is absolutely essential for all managers. Currently, the discussion in books, articles, and lectures is heavily focused on the technical aspects of this fourth industrial revolution. In this book, the author places emphasis on the management perspective. Managers are ultimately the decision-makers but also the drivers of the necessary changes toward a digital and smart enterprise. This transformation also—and especially—includes corporate culture. It is therefore not sufficient to simply acquire a few sensitive robots, autonomous transport vehicles, or 3D printers for one's own company. The required transformation and change must go much deeper—it must encompass the entire organization. However, the decisive success factor for implementation and thus the successful introduction of Industry 4.0 lies outside of technology. The qualification of all employees in the company is of central importance. It is not enough to offer corresponding web-based training. What is needed are more comprehensive and significantly "smarter" solutions, such as the use

W. Huber, *Industry 4.0 Compact—How Technologies Are Transforming Our Economy and Our Companies*, https://doi.org/10.1007/978-3-658-51029-9_1

of gamification and innovative learning formats (for example, “knowledge to go” in the form of short learning units).

Over the past more than 10 years, it has become clear that Industry 4.0 and digitalization are not just buzzwords invented by creative marketing departments, but represent a real trend toward another profound transformation of our economic world—that is, how and what we will produce in the future and thus secure our prosperity. This change is not limited solely to manufacturing and production, but affects all areas of the company: from product development to aftersales and all indirect departments. It is therefore not only a technical revolution, but above all a management revolution. As is often the case, technical aspects can be mastered and changed much more easily and quickly than employees, corresponding organizational structures, and their culture.

As a consequence of digitalization and Industry 4.0, completely restructured value chains are emerging, leading to a reorganization of our entire economy. How do companies interact with each other, what is the significance of partnerships, and how must companies and the economy adapt to these changes? All these questions are addressed. This also includes a look at the international landscape, with both old and new challengers for our economy.

To actively shape these changes, compact and up-to-date knowledge of the relevant technological drivers, products, and solutions is required. In this book, the key players are presented in a focused manner in the form of individual technologies. The impact on business models and, consequently, on how “smart,” i.e., intelligent products, are changing the economy, constitutes an important part of this book. The associated changes to business models and products are thus also incorporated into the practical presentation. The individual topics are illustrated with clear examples. The risks as well as opportunities associated with “new” products and production methods are presented in a practice-oriented manner.

However, Industry 4.0 and digitalization also have a significant impact on indirect areas such as controlling and human resources, and ultimately on how employees and companies will be managed and decisions made in the future.

Drivers of Change

2

Companies, regardless of industry, are currently facing enormous challenges. Managing crises and uncertainties has become a key task, and thus so has dealing with risks. Digitalization can provide support in this regard, but at the same time, it is also a driver of change. This reveals its Janus-faced character. Digitalization is therefore transforming our daily lives in practically all areas, both private and professional.

This digital transformation is not an abstract phenomenon, but can be very well illustrated visually, see Fig. 2.1, see [12], [13]. Expressed somewhat more abstractly in numbers: In 2015, around 80% of German citizens had access to the internet, 63% of whom used it daily. A large proportion of users accessed the internet via mobile devices. 9 out of 10 Germans made purchases online in the same year, see [1]. By 2022, the number of internet users had risen to 95%. In the 14 to 49 age group, the proportion was even 100%. Even in the 50–69 age group, the share was very high at 95%, see [9]. Companies that do not participate digitally in the future will sooner or later be pushed out of the market. The task now is to further develop this mindset.

Since the "invention" of the World Wide Web in 1987, a great deal has happened. Within just a few decades, user behavior has changed dramatically, if not completely. This pressure for change will continue to increase in the coming years. In 1993, around 14 million people worldwide used the internet, and by 2013, this number had already risen to 2.7 billion people, about one third of the world's population, see [7]. In 2021, the global share of internet users rose to around 4.9 billion, or 66.2%. By 2022, there were already over 5.2 billion people online worldwide. Numerically, China leads the list, followed by Europe, with North America in third place, see [8].

Companies must master the digital transformation towards Industry 4.0. For Germany, as a leading industrial nation, it is essential to successfully shape this transformation. It is therefore of central importance to understand exactly what digitalization and Industry 4.0 mean and how they can be leveraged. Opportunities as well as risks must be

W. Huber, *Industry 4.0 Compact—How Technologies Are Transforming Our Economy and Our Companies*, https://doi.org/10.1007/978-3-658-51029-9_2

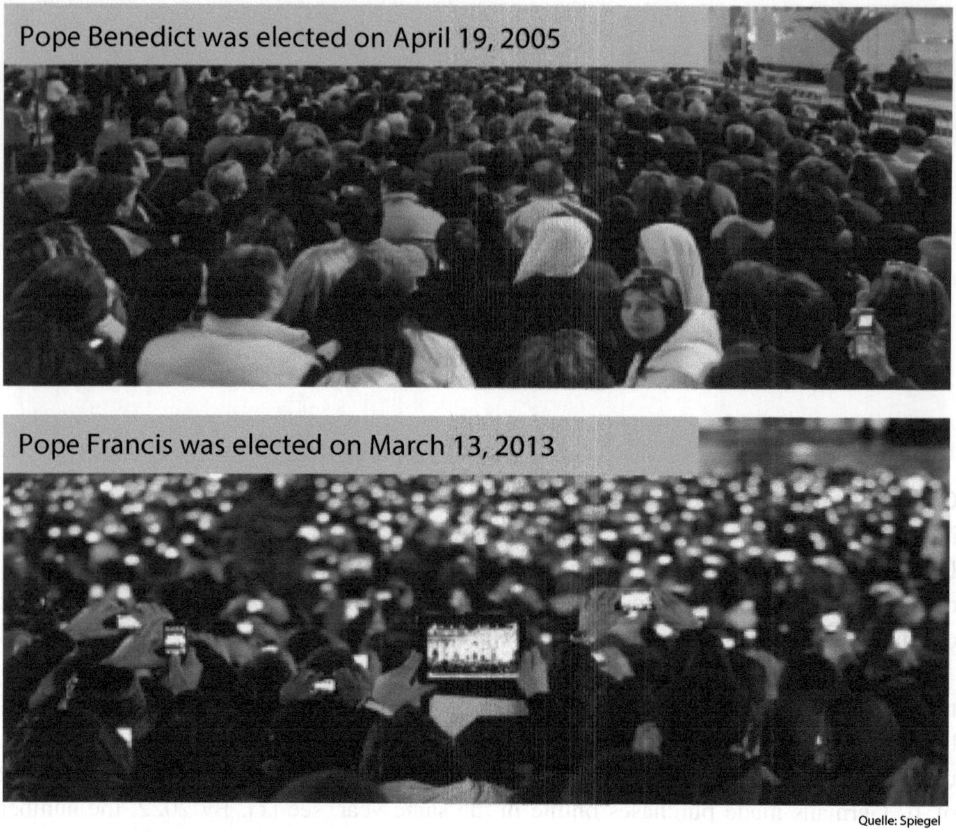

Fig. 2.1 Changed consumer behavior

identified, with the aim of seizing the opportunities while minimizing the risks. Companies that cannot keep pace with rapid developments will quickly jeopardize their market position. However, digitalization and Industry 4.0 do not mean that everything technically possible should be implemented, but rather that actions should be taken with sound judgment and economic sense. The focus is not solely on generating economic benefits, but also on the management of uncertainties and risks, as mentioned at the beginning of the chapter. Uncertainties are manifested very practically in highly volatile consumer behavior and thus in highly fluctuating delivery call-offs, for example, from automotive manufacturers to their suppliers.

So what are the technical drivers of these profound—revolutionary—changes, and why now?

One of the key insights is that data will play an increasingly dominant role for companies. Data is the proverbial new raw material. A McKinsey study reached a similar conclusion. According to this study, as early as 2015, international data flows contributed more to global economic growth than traditional goods trade, see [8], [10]. Similar results are evident in the transformation of markets, which are increasingly data-driven, see [14].

In general, the changes discussed result from so-called "mega-trends" and "game changers" (the latter are characterized by having a significant impact on companies and their business practices). The aforementioned software and the associated data are such a game changer, see [1].

The significance of mega-trends is strongly industry-specific. For example, the automotive industry is subject to much stronger changes than, say, the aerospace industry. In Fig. 2.2, mega-trends are illustrated with a focus on the automotive industry as an example.

Naturally, this perspective in detail also depends on the individual company and thus its market positioning and market segment. A volume manufacturer in the automotive sector, such as VW, will certainly view these "mega-trends" and their individual significance somewhat differently than, for example, BMW as a representative of the premium segment. The same applies analogously to suppliers. Aerospace companies will draw their own conclusions from the mega-trends.

To what extent the trend toward individualization can also be considered a mega-trend is left to the reader. Regardless of personal assessment, it is an important driver—not only in the consumer sector. In general, product life cycles will drastically shorten, with significant impacts on companies. Formerly long product life cycles led to customized and thus optimized machines and organizations. Short product life cycles, on the other hand, require modular, flexible, and scalable manufacturing and production structures, including indirect areas and thus the entire organization. The already repeatedly mentioned management of uncertainties and risks is certainly among the new mega-trends.

As an example of mega-trends, demographic change is illustrated, see Fig. 2.3.

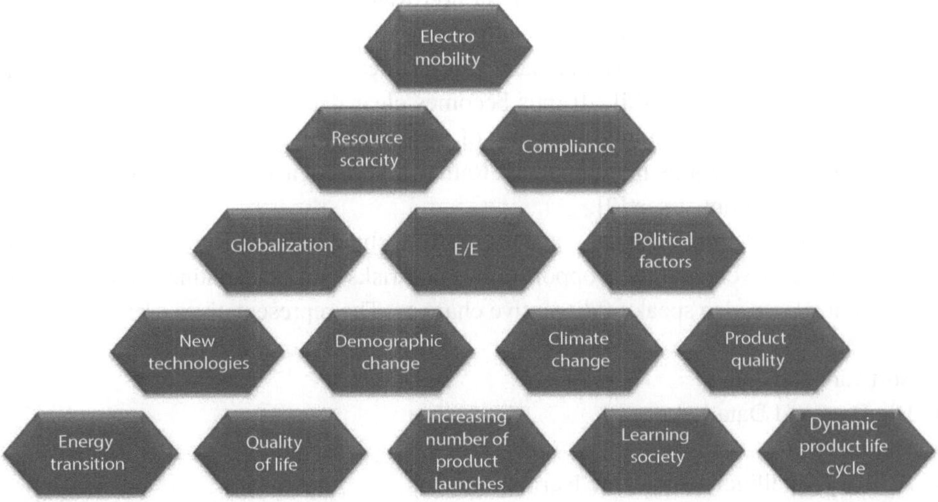

Fig. 2.2 Mega-trends from the perspective of the German automotive industry

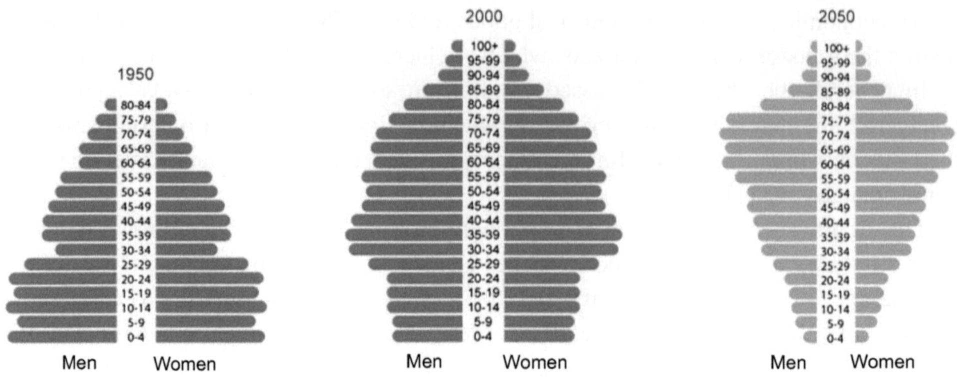

Source: centerstage.de, based on Federal Statistical Office, 2016

Fig. 2.3 Demographic change

The proportion of German citizens younger than 65 will decrease significantly in the coming decades. This will have a considerable impact on the world of work. Working conditions for employees "65+" will often need to be created or significantly expanded. It should be noted that this change will have very different effects geographically. This will also be accompanied by a change in consumer behavior and thus a change in domestic demand. The transformation is associated with increased demand for qualified employees (both academics and skilled workers) and a decline in the number of low-skilled workers. This results in a greater need for qualification measures, not only for the low-skilled, and a need for increased permeability in training and career models. Another point is maintaining and expanding the employability of a company's own employees. To what extent demographic change will affect non-wage labor costs and thus contributions to social security is still open. The steadily increasing degree of automation must also be taken into account. Furthermore, new job profiles will be created, and the demand for existing personnel will shift. It thus becomes clear that mega-trends will have a significant impact on the economic structure of this country. As a result, technologies as a whole will not reduce jobs in companies. Rather, technologies contribute to maintaining prosperity in our country, see [7].

The so-called "game changers" are primarily technological drivers. In the following chapters, these drivers and their opportunities and risks will be explained in more detail. In this context, we also speak of disruptive changes. The representatives are:

- Software and data
- Big Data and Data Lake
- Digital twin
- Artificial intelligence/machine learning
- Cyber-physical systems

- Sensitive robots
- Additive manufacturing with 3D printers
- Platforms

These technologies therefore have revolutionary potential. However, there are also technologies with a distinctly evolutionary character. These include:

- Cyber security
- Vertical and horizontal integration
- Blockchain
- Autonomous transport units
- Real time enterprise and mobile solutions
- Digital factory
- Assistance systems
- Cloud
- 5G/6G
- Metaverse
- Quantum computing

The individual technologies (regardless of whether revolutionary or evolutionary) are the technological component of Industry 4.0 or the Internet of Things (referred to as the Internet of Things in American English). To highlight the added value for all stakeholders, the Internet is further expanded to include services (Internet of Services).

Industry 4.0 is a German invention. This coined term was presented at the Hannover Messe in 2011. A detailed description of the term will follow in the next chapter.

The topic of cyber security and security in production (OT security) occupies a somewhat special position here. Its importance increases with the use of the technologies mentioned here, but it is also the result of the current crisis situation. Therefore, it is also classified among the revolutionary technologies.

Naturally, the technologies mentioned above (regardless of whether they are revolutionary or evolutionary) have varying degrees of maturity in terms of their practical applicability. In Fig. 2.4, the maturity level of the individual technologies is shown as an example from Gartner's perspective.

The technologies discussed here can be seen as challenges, but also as possible solutions to address the mega-trends—again, both risks and opportunities. As already mentioned, dealing with uncertainties and risks as another mega-trend is not a technical issue. However, it has significant impacts in this area and will therefore also be considered separately at the end of the chapter, since technologies can support and thus be or become part of the solution.

Another mega-trend already mentioned is sustainability. Thus, in addition to digital transformation, an ecological transformation must also be managed. However, the two are not mutually exclusive, but rather interdependent. Through consistent digitalization,

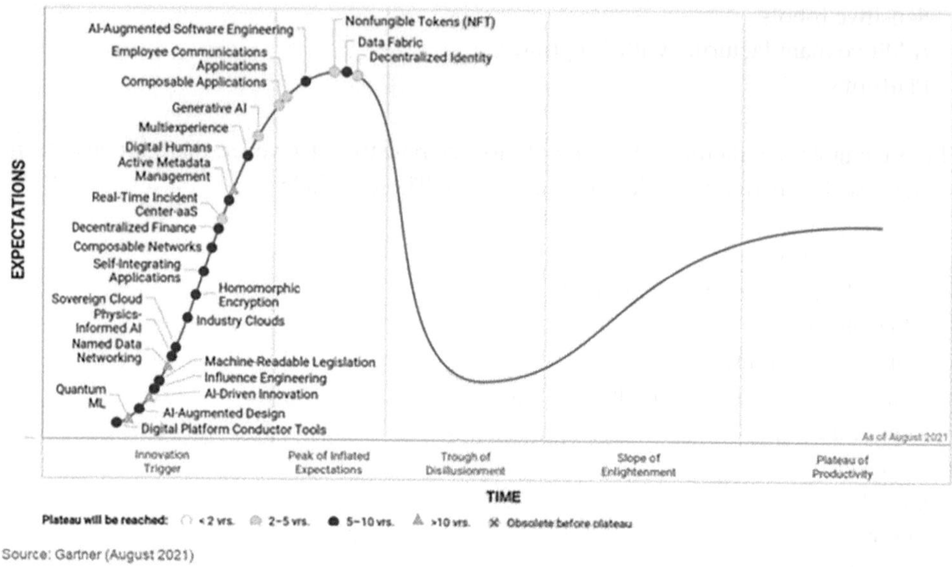

Fig. 2.4 Maturity level of Industry 4.0 relevant technologies

approximately 17% of the required CO2 savings can be achieved and up to 50% of total resource consumption can be reduced, including a reduction in material requirements, see [4, 15].

But how exactly do technologies change companies?
Digitalization and Industry 4.0 change processes. Processes are therefore not just optimized. When processes are changed, we speak of digital transformation, to differentiate it from a more or less 1:1 mapping of manual processes into digital ones. In the following, this differentiation will be dropped, except where explicitly stated.

In addition, possibilities and functionalities arise that would not have been possible without the corresponding technologies. They determine and drive business success. New business models are thus created, see [1]. The development and expansion of new customer relationships and digital ecosystems is possible, and at enormous speed. Digitalization thus changes value chains and reduces market entry barriers. Ultimately, established companies are existentially threatened. Nokia and Agfa are concrete examples—victims—of this digital disruption, i.e., change. Both companies failed to recognize the changes or recognized them too late and failed. This list can be extended at will and, unfortunately, continues to grow.

However, digitalization and digital transformation also open up numerous new opportunities for companies to expand their existing business models. Traditionally structured companies with a clear organizational structure struggle to take advantage of these opportunities. Rigid departmental and divisional boundaries prevent this. Cross-functional thinking and working in heterogeneous teams create the necessary opportunities and freedom here.

Industrial development has also shown that job profiles disappear due to technical innovations and new ones emerge. Traditional industrial regions, such as Germany's Ruhr area or Manchester in England, have lost significant economic importance due to changing conditions and technological progress, or have had to "reinvent" themselves. In some cases, they even became economic and social problem areas. These examples from the past show that "sleeping through" changes (whether due to mega-trends or game changers—i.e., technologies) poses a significant risk, not only for companies but also for the associated societies and economies.

Doesn't your own company still have time to change and learn from the experiences of others?
The examples at the beginning of the chapter have hopefully illustrated that changes brought about by digitalization happen very quickly. This applies not only to the consumer sector, but is a paradigm, as shown in Fig. 2.5, see [5], [11]. The dynamics of change can also be seen in companies. Since 2000, around 50% of the world's largest companies by revenue (Fortune 500) have disappeared. In addition, digital companies generate the highest profit margins in the world, see [4].

The US hearing aid industry is a very illustrative example, even though it is certainly more of a niche market. Within less than two years, the entire industry switched from conventional to additive manufacturing, see [6]. Thus, the credo is: Either you change, or you will be changed. Further examples of changing industries and companies can be found in the following chapters. This is also consistent with a study by Capgemini, which sees an average response time of 18 months for traditional industries in the face of disruptive changes, see [3]. Of course, this period varies from industry to industry and from market segment to market segment. Unfortunately, however, it does not change the fundamental trend.

This is also illustrated by a study conducted by the consulting firm Roland Berger, see Fig. 2.6. Here, the potential for losses was examined. It shows the considerable threat potential that could arise for companies from digitalization and digital transformation. Over time, these forecasts have proven to be accurate.

Technology	Period to gain 50 million users, stated in years
Radio	38
Television	13
Facebook	3.5
Cell phone	1
Twitter	0.75
WhatsApp	0.15

Fig. 2.5 Time taken to reach 50 million users

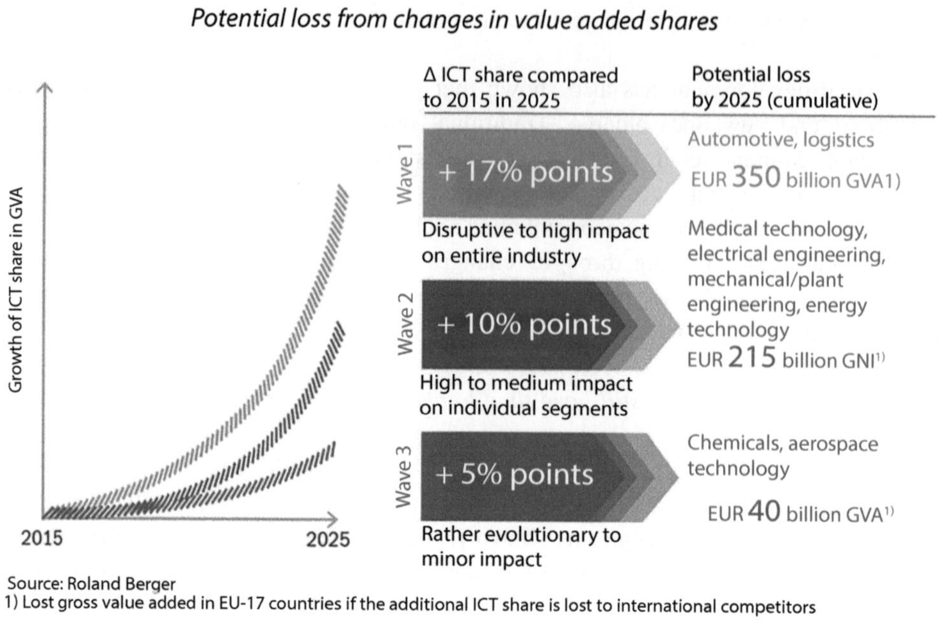

Fig. 2.6 Impact of Industry 4.0 on value creation

To counter these potential losses, it is unfortunately not enough to optimize only one's own production in the form of a so-called smart factory; instead, intelligent and networked—i.e., smart—products and novel value creation processes, and thus business models, must be implemented at the same time.

Many companies have set out in recent years and implemented corresponding strategies and projects. However, this is unfortunately not true for all companies. Around a quarter of companies are still more or less operating in an analog manner. Thus, a lot has changed for the better in Germany in recent years. Unfortunately, the German Mittelstand is lagging somewhat behind in this area, which can have fatal consequences for Germany in particular, since the Mittelstand is the backbone of our industry and thus our prosperity. However, it is also evident that digitalization makes companies significantly more sustainable.

This lack of a comprehensive digital strategy has also been identified by the market research institute IDC in a study, see [2]. The study does not represent an isolated opinion. There are numerous investigations and studies on this topic, both from the academic and non-academic sectors, such as consulting firms.

Digitalizing one's own company therefore means much more than simply operating a website or engaging in electronic commerce.

Digitalization thus changes existing business models and requires fundamentally new thinking—especially at the executive level, see [1, 4]. This changed way of thinking, however, means a transformation of the company. In other words, a digital mindset is required. More on this in the next chapter.

Smart products are already fundamentally changing the competitive situation in many industries. When used correctly, they create the opportunity to achieve stronger market and product differentiation. In addition, data-based services, both for end customers and for the company itself, further reinforce this trend.

If these new products are combined with correspondingly new digital business models and a smart factory for their production, the revolutionary character becomes apparent. At the end of this development are highly individualized products designed to maximize subjective/individual customer benefit. This requires production without fixed cycles and assembly lines, in the form of adaptable and flexible systems and structures. Examples of this are provided by innovative approaches from RWTH Aachen in cooperation with Porsche.

On the other hand, employees expect modern workplaces and infrastructures, innovative companies that can also respond flexibly to changing life situations, and provide space for personal development and thus further advancement. Hierarchical corporate structures and centralized decision-making processes with slow speeds are seen as rather off-putting. New employees expect significantly more decision-making authority, but also increasingly the compatibility of work and family or private life.

To realize the resulting potential, however, changed corporate structures are also required. The main obstacle to digital transformation is the defense of existing structures (both internal to the company and in market structures) combined with a lack of willingness to change and thus lifelong learning. This is also associated with new, i.e., digital customer relationships and digital ecosystems. These can be used to build market entry barriers against potential competitors. A tangible example in the consumer sector is Apple. Switching to another provider leads to a significant loss of functionality/comfort, as the services are not compatible with each other and data or content transfer is not the easiest task.

Ultimately, price and efficiency in manufacturing lose significance. If a product is no longer in demand on the market due to changing customer and market needs (see megatrends), then the price and efficiency of its production are irrelevant. This results in a shift in importance from production to (smart) product development.

To achieve this, more and more platforms are being established, virtually as "intermediaries" between manufacturing companies and end customers. These platforms at least temporarily create added value for end customers. This platform concept is becoming established not only in the B2C but also in the B2B environment. Regardless of the respective benefits, a combined product and business model disruption is thus emerging. The goal is to be the "first" to bring new ideas to market. According to a Bitkom study, about half of the companies surveyed use digital marketplaces, with a strong upward trend. However, these platforms are also threatened by technological developments (the revolution devours its children) in the form of matching algorithms. These will be the future core of such platforms. They map customer needs to the products offered. It sounds simple, but is in fact highly complex, see [14].

References

1. Manfred Broy, Christian Prehofer, Hermann Engesser, Das Interview: Digitalisierung und die Rolle der Informatik in der Anwendung und Forschung, Informatik Spektrum Band 39, Heft 6, 12/2016.
2. IDC Studie: Industrie 4.0 durchdringt verarbeitendes Gewerbe in Deutschland, 2014.
3. Industrie 4.0 – Eine Einschätzung von Capgemini Consulting, 2014.
4. Jürgen Hill, Studie zu Industrie 4.0: So digital ist Deutschlands Industrie, Computerwoche, 25.05.2022.
5. Tobias Kollmann, Holger Schmidt, Deutschland 4.0: Wie die Digitale Transformation gelingt, Springer Gabler, 2016.
6. Ralf T. Kreutzer, Karl-Heinz Land, Digitaler Darwinismus: Der stille Angriff auf Ihr Geschäftsmodell und Ihre Marke, Springer Gabler, Februar 2016.
7. Manuela Lenzen, Künstliche Intelligenz, C.H. Beck, 2018.
8. L. Lohmeier, Anzahl der Internetnutzer weltweit bis 2022, statista, 05.05.2023.
9. L. Lohmaier, Statistiken zur Internetnutzung in Deutschland, statista, 31.08.2023.
10. Michael Porter, James E. Heppelmann, Wie Smarte Produkte den Wettbewerb verändern, Harvard Business Manager, 12/2014.
11. Armin Mahler, Thomas Schulz, Digitale Revolution – Wie die Internet-Ökonomie unser Leben verändert, Spiegel-Verlag, 2014.
12. Jürgen Meffert, Heribert Meffert, Eins oder Null: Wie Sie Ihr Unternehmen mit Digital@Scale in die digitale Zukunft führen, Econ, 2017.
13. Ulz, Digitale Erleuchtung, Spiegel Online. http://www.spiegel.de/panorama/papst-momente-bilder-zeigen-vergleich-zwischen-2005-und-2013-a-889031.html. aufgerufen am 11.04.2018.
14. Viktor Mayer-Schönberger, Thomas Ramge, Das Digital – Markt, Wertschöpfung und Gerechtigkeit im Datenkapitalismus, Econ Verlag, 2017.
15. Studie: Ressourceneffizienz durch Industrie 4.0 – Potenziale für KMU des verarbeitenden Gewerbes, VDI, Juni 2017.

3 What Is It About? Definition and Classification

The individual industrial revolutions, from the first with the introduction of mechanical production facilities to the present day—the fourth industrial revolution—have had, or are having, a significant impact on the prosperity of individual nations or population groups, see Fig. 3.1. This insight is therefore not new and should not really come as a surprise with the fourth industrial revolution. Occasionally, there is already talk of Industry 5.0, with a focus on networking, data management, and comprehensive and holistic application of technologies, see [10]. There is a stronger focus on a human-centered and fully integrative approach. Upon closer examination, however, all these elements are already included in the Industry 4.0 approach or should be taken into account. If you look at practical implementation examples of Industry 4.0 projects, these elements are already in focus, especially in the factory and product development, see also the examples in the following chapters. Thus, it is up to the reader whether to speak of Industry 4.0 or 5.0. In the further course of the book, we will therefore retain the terminology Industry 4.0.

With each stage of development, the value creation by humans also changes. In the first industrial revolution, humans were still the operators of machines; in the second, they became specialists; and in the third, enablers. In the fourth stage, humans are now considered designers—not only of machines, but of the entire value chain. This perspective also contradicts a further stage of development toward 5.0, at least in the current definition, see [10], which claims this aspect for itself. That there will certainly be further stages of development after Industry 4.0 is undisputed. What these will look like is currently still unknown.

Industry 4.0 encompasses all industrial areas of a company, from product development to sales and aftersales, see Fig. 3.2.

W. Huber, *Industry 4.0 Compact—How Technologies Are Transforming Our Economy and Our Companies*, https://doi.org/10.1007/978-3-658-51029-9_3

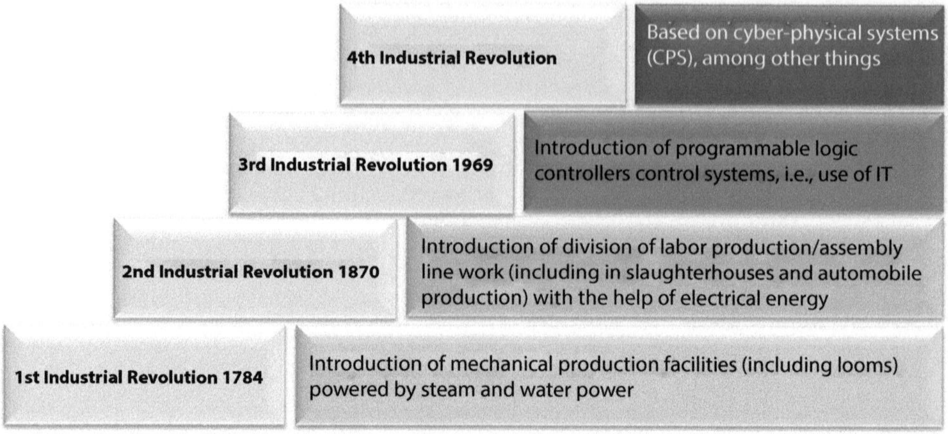

Fig. 3.1 The four industrial revolutions

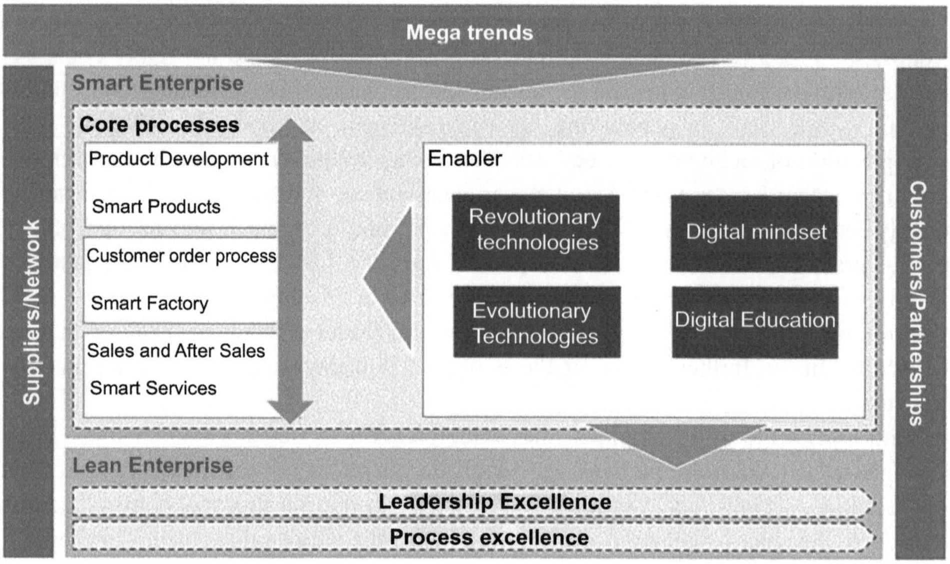

Fig. 3.2 Definition of Industry 4.0

The focus of Industry 4.0 is clearly on the core processes of a company, regardless of the respective industry:

- Product development: Smart Products,
- Customer order process: Smart Factory,
- Sales and aftersales: Smart Services.

Supporting processes, such as procurement or controlling, are also subject to change, but are not mentioned separately. In these areas, AI (artificial intelligence) can, for example, take over routine tasks. However, digitalization and digital transformation have also fundamentally changed these areas. The focus, however, is on the value-adding and differentiating core processes of a company.

In addition, a company transformation toward a smart/digital enterprise with new forms of leadership is required, i.e., digital leadership or Management 4.0. In recent years, practical experience has shown that digital education and a digital mindset are essential for change. The enablers of change are technologies. The already mentioned megatrends are the drivers of change. The individual topics will be examined in more detail in the following chapters.

The digital mindset is about transforming the entire company and its employees to use data and algorithms (including artificial intelligence) to create new business opportunities and optimizations/improvements. The technical prerequisite for this is easily accessible data for the individual departments. The departments know their specific challenges and can derive corresponding use cases and implement them. Overall, this represents a paradigm shift away from specific solutions by IT departments or external consultants or solution providers toward applications by the departments themselves, see [8].

In short, Industry 4.0 (I40) is the linking of intelligent products (smart products) with an intelligent factory and production (smart factory) and changed business models (smart services). The basis for this is the easily available, high-quality data already mentioned in the previous section. In sum, it is therefore an integrated and thus holistic approach.

In the context of Industry 4.0, there are many terms and supposed synonyms. In Fig. 3.3, various frequently used terms are set in relation to each other, see also [5].

Here, digitalization is a necessary prerequisite and an integral part of Industry 4.0. Unfortunately, digitalization is often equated with Industry 4.0. Digitalization refers to the transformation toward a process-driven and paperless company using hardware and software. This means networking all areas of a company. As a result, data and software become important tools for companies and managers. This inevitably includes collecting and using company-relevant information and translating it into concrete actions, see [9]. Industry 4.0 goes far beyond this approach, as shown in Fig. 3.2 and the above description. Thus, the term "smart" is explained as the combination of lean processes, intelligent automation solutions, and digitalization. This results in a digital transformation of existing processes across all areas of a company. By combining the three elements—processes, automation, and digitalization—new opportunities and thus optimization potentials arise. The term digital transformation was already briefly explained in the previous Chap. 2. Thus, the result of a digital transformation of all production processes is, for example, a smart factory.

Digitalization and Industry 4.0 have now reached all industries, such as the automotive industry and mechanical engineering, see [2]. This is also evident in the fact that the "digital economy" in Germany reached a volume of €73 billion in 2015, which corresponds to around 3% of the gross domestic product (GDP). In 2021, it was already 5.7%, although

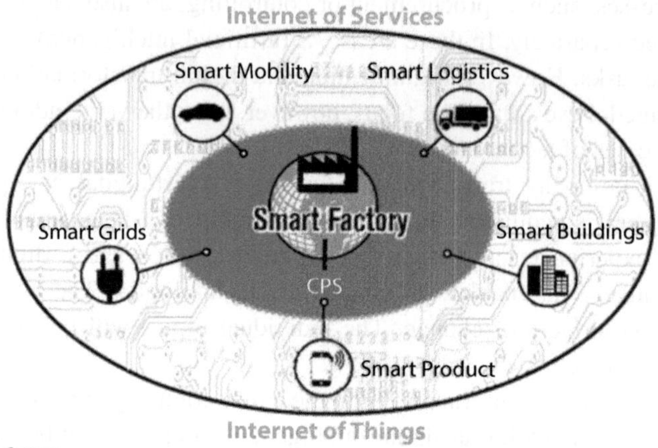

Source:
Final report on Industry 4.0

Fig. 3.3 Classification of the various terms

still significantly less than in the USA at 8.2%. However, this representation does not take into account industries such as the automotive industry, as mixed products are the focus here. If these are also included in the calculation, the share in Germany amounts to approx. 13%, analogously for the USA, see [4]. This trend is also visible in the labor market. Since 2017, the IT sector, with over 1 million employees, has been the largest industrial employer in Germany, see [1]. Digital technologies have become established and are ubiquitous. For example, in the same year, we Germans spent around 2.6 h per day on the Internet, see [7]. In 2021, it was already over 70 h per week. This increase is certainly partly attributable to the pandemic, but it shows a clear upward trend, see [11]. The dynamics of Industry 4.0 and digitalization will continue to advance, see [7].

According to Joe Kaeser (former CEO of Siemens), two factors determine the pace of digital transformation: digital performance and creativity. According to Joe Kaeser, digital transformation heralds a paradigm shift in the economy, see [3].

In general, neither digitalization nor Industry 4.0 is primarily an IT issue. An effective and efficient IT infrastructure is solely the foundation upon which both initiatives are built. The goal, in addition to developing new business models, is their efficient execution. In the digital transformation towards an Industry 4.0 enterprise, the technologies briefly mentioned in the previous chapter are utilized to develop the company. Since all core and support processes are affected, digitalization changes the entire structure, organization, processes, and ultimately the people within a company. Therefore, the respective specialist department should be the driving force behind this change.

Management must explicitly not be excluded from this process. The changes brought about by digital transformation are most profound for them. Leadership behavior, employee communication, and the required skills associated with these also change significantly. Managers must make decisions more quickly and, among other things, use and ultimately master communication with modern digital media. Consequently, corporate structures and their culture also change.

As a result, the need for and the manner in which employees are and should be qualified also changes. The motto "lifelong learning" should therefore not be seen as an empty phrase, but rather as a paradigm. More on this in later chapters.

Overall, it is about the transformation of products, production, service/aftersales, corporate organization, and management processes. Consequently, it would be more appropriate to speak of Economy 4.0 rather than Industry 4.0 here, see [6].

References

1. Martin Bayer, Deutsche ITK-Branche wird zum größten industriellen Arbeitgeber, Computerwoche, 01.11.2017.
2. Manfred Broy, Christian Prehofer, Hermann Engesser, Das Interview: Digitalisierung und die Rolle der Informatik in der Anwendung und Forschung, Informatik Spektrum Band 39, Heft 6, 12/2016.
3. Joe Kaeser, From Data to Business: Neue Geschäftsmodelle deutscher Industrieunternehmen, in Thomas Becker, Carsten Knop (Hrsg.), Digitales Neuland: Warum Deutschlands Manager Revolutionär werden, Springer Gabler, 10/2015.
4. Martin Greive, Deutschlands digitale Wertschöpfung laut IWF-Studie „verbesserungsfähig“ Handelsblatt, 14.03.2021
5. Henning, Kagermann, Wolfgang Wahlster, Johannes Helbig, Umsetzungsempfehlung für das Zukunftsprojekt Industrie 4.0 – Abschlussbericht des Arbeitskreises Industrie 4.0, 4/2013.
6. Ralf T. Kreutzer, Treiber und Hintergründe der digitalen Transformation in D. Schallmo (Hrsg.), Digitale Transformation von Geschäftsmodellen, Springer-Verlag, 2017, Seite 33–58.
7. Tobias Kollmann, Holger Schmidt, Deutschland 4.0: Wie die Digitale Transformation gelingt, Springer Gabler, 2016.
8. Tsedal Neeley, Paul M. Leonardi, Wie Ihre Mitarbeiter ein digitales Mindset entwickeln, Harvard Business Manager, August 2022.
9. Daniel R. A. Schallmo, Jetzt digital transformieren: So gelingt die Digitale Transformation Ihres Geschäftsmodells (essentials), Springer Gabler, 2016.
10. www.industrie-50.de/definition. aufgerufen am 26.11.2022.
11. Dpa, Jugendliche mehr als 70 Stunden pro Woche online, Zeit online, 12. August 2021.

and ultimately master communication with modern digital media. Consequently, corporate structures and their culture also change.

As a result, the need for the manner in which employees are and should be qualified also changes. The motto "lifelong learning" should therefore not be seen as an empty phrase, but rather as a paradigm. More on this in later chapters.

Overall, it is about the transformation of products and production [illegible]

References

[illegible]

4 Technologies That Revolutionize

In Chap. 2 those technologies with revolutionary potential were already introduced. These game changers, and why it is precisely the ones listed, are the subject of this chapter. But first, a brief explanation and motivation for the distinction between technologies of a revolutionary and evolutionary nature will be provided.

First, regarding the term game changer. In our context, this Anglicism refers to technologies that fundamentally alter existing processes, strategies, and companies. The technology thus brings about a paradigm shift, a groundbreaking innovation, upheaval, or reorganization. This translation from various translation programs captures the essence very accurately.

An alternative term here is disruptive, as in disruptive technology or disruptive business models. Here, too, an existing product or service is completely redefined through digitalization and Industry 4.0. Existing providers may be replaced by new ones, and technical components lose their significance. A well-known example is photography. About 20 years ago, you needed a camera and film, had to develop prints, and so on. Today, a smartphone, which among other things has a camera, is sufficient, and personal experiences can be shared instantly with others via social networks, without time delay and usually without additional costs, see [30].

Thus, the terms "game changer," "disruptive," and "revolutionary" are treated as synonyms. So much for the terminology. The classification of individual technologies as game changers is somewhat more difficult. It is partly subjective. Therefore, the designation of the technologies listed here as game changers is ultimately subjective. Nevertheless, for better comprehensibility, a more in-depth explanation and motivation are provided. It should also come as no surprise that these game changers can vary from industry to industry.

Regardless of the classification, individual technologies should not be viewed in isolation. This is also one of the important points in the following chapters. It is precisely

W. Huber, *Industry 4.0 Compact—How Technologies Are Transforming Our Economy and Our Companies*, https://doi.org/10.1007/978-3-658-51029-9_4

the combination, connection, and ultimately the networking of individual technologies that creates significant added value—synergies, in other words. This chapter will hopefully also make it clear that, strictly speaking, all technologies only unleash their full explosive power when they interact. This networked character is particularly evident in the first four technologies discussed in this chapter.

However, it is important to correctly assess and evaluate the opportunities and risks of new technologies. In addition, the consequences for management must also be discussed, regardless of whether the technologies are disruptive or not. Missing a technological trend usually has serious consequences for a company's economic success. With disruptive technologies, the negative economic effect is usually much greater. This applies to all areas. It is particularly severe in product development, production, and sales & after-sales (i.e., with new business models), thus in the core processes of companies.

4.1 Data, Software, and Smart Data

Data is often referred to as the new raw material or as the fourth factor of production (after labor, capital, and raw materials). But what exactly is data? What distinguishes it from smart data, or in other words, how does data become smart data?

4.1.1 Technical Description

Everything around us is data, regardless of whether it is useful to us or not, whether it is meaningful or not. Strictly speaking, data is a collection of symbols and, at first, has no meaning or sense. Only by establishing relationships between these symbols do they acquire meaning. The data thus becomes significant for us. Data thus becomes information. Only the processing of information results in knowledge, or in modern terms, smart data. This processing or transformation (a term preferred by computer scientists) of data can be performed either by our human brain or by software. Ultimately, the goal is to derive actions in order to generate the desired added value. The latter (i.e., machine processing) has the great advantage that much larger volumes of data can be processed (i.e., transformed), and at a much higher speed than by the human brain. This is illustrated again in Fig. 4.1. An example of processing large volumes of data is AI-based approaches. We will discuss this shortly.

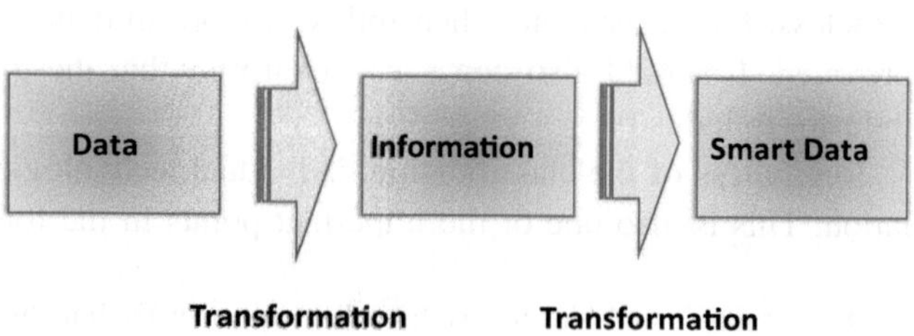

Fig. 4.1 Transformation from data to smart data

To express the generated data flood in numbers: around 2.5 exabytes of data are created every day (as of 2012), see [23, 45]. In 2018, it was already 33 zettabytes. Of this, the manufacturing sector accounted for 3.6 zettabytes or 3600 exabytes, see [43]. By 2025, this value is expected to grow to 175 zettabytes, about 8 times as much as in 2017, see [72]. Even if this data flood only marginally affects us personally or professionally, a considerable volume of data remains, and it is further complicated by the fact that this data bombards us in various forms (emails, PowerPoint presentations, images, Excel sheets, etc.). The much-vaunted gut feeling finally loses its justification with these large volumes of data, and we are slowly becoming overwhelmed. This enormous flood of data can no longer be efficiently processed and managed in everyday professional life without software. We will discuss the technology for managing data in the form of big data separately in this chapter. So, a little patience—first, the basics.

Now we come to the second term—software. Industrial data processing is not possible without software. Alongside data, software significantly shapes our lives. Without both, we would not have, for example, smartphones. The software-supported processing of data has a significant impact on the economic fortunes of every company, see [1]. It even gives rise to new business models. Software and data thus take on a dominant role—a strategic tool—in successful companies. So it is not surprising that data-driven companies are about 5% more productive and about 6% more profitable than their competitors, see [23]. Thus, processes change based on the available data. The combination of software and hardware leads us to so-called cyber-physical systems (CPS), which will also be discussed further below. The aforementioned smartphones, by the way, are an example of CPS in everyday life. So, we will encounter software in many different forms and manifestations.

4.1.2 Consequences for Management

Data and software are becoming increasingly important, regardless of company size or industry. Companies must therefore build or expand their competence in this area and make use of this data. This is not limited to IT departments. Above all, expertise in handling data must grow within the business units, as this is where economic success is generated. These units must also drive these topics forward. Consistent use of data (both one's own and customer data) creates new business areas and thus opportunities; more on this in Chaps. 6 and 8. It is also possible to improve internal business processes, such as in production. We will address this in Chap. 7. Management is called upon to make the necessary strategic decisions here. This can take the form of spin-offs, such as start-ups (as seen with Kuka or Trumpf), or through offices and branches (as practiced by Audi), or by establishing new divisions. Regardless of the chosen approach, market dynamics should not be underestimated. Functionality in the form of software can be developed much faster than in hardware, while offering significantly greater flexibility, for example through updates or dynamic feature expansion (at John Deere, the required engine power can be controlled via software, and Tesla demonstrates what functionality can be added via software even after a car has been purchased, such as semi-autonomous driving). BMW is taking a similar path. Since 2018, models can be supplied with updates for life, including additional functionality. BMW refers to this as the adaptive new car, see [11]. The replication of software occurs at virtually zero cost, which is also an important point.

The greatest challenge for companies and thus for management is overcoming their own inertia, as well as building the necessary competencies (including at the top management level). This requires overcoming internal barriers. This also includes changing one's own mindset, that is, the shift toward a dynamic and learning organization, which ultimately also involves changing the way work is done—and, as is well known, this can be changed without altering the organization itself (at least to some extent). The digital mindset required for this was already discussed at the beginning of the book.

The use of data can be divided into three maturity or development stages:

- Stage 1: Building data competence,
- Stage 2: Using data,
- Stage 3: Data-driven business models.

In the first stage, data competence is built within the company, that is, awareness of the value of data is created. This forms the basis for development and thus for utilization. Systems, which form the basis for data, are not yet fully integrated at this stage. As a result, data may exist multiple times, in different forms (an order may be represented differently in SAP than in a custom-developed system), and with different temporal granularity. In the second stage, data is used much more consciously. Only in the third stage does the full transformation take place, with the use of data-driven business models (more on this in Chap. 8). This also involves the transformation of management, as already mentioned.

It is evident that the conscious and controlled use of data has many similarities with the topic of digitalization. Here, software-based value-added services are also deployed comprehensively and strategically. As a result, products are becoming increasingly digitalized. However, direction and control must be set by top management. In general, the rule is: everything that can be digitalized will be digitalized—and (almost) everything can be digitalized.

With the maturity and thus more intensive use of data—not only one's own, but also that of customers—the question quickly arises as to who owns this customer data. Whoever controls the data controls the customers and thus the market. In many cases, one knows the customer better than the customer knows themselves. US companies such as Google and Amazon excel at this business. German automotive manufacturers have already made the shift in perspective toward data utilization or are in the process of doing so. Other industries are still lagging somewhat behind this trend. German companies must move away from the product perspective in favor of the service perspective.

This point will run like a "red thread" or mantra throughout the chapter and the entire book. This is an uncomfortable truth, as the product business still often yields good to very good margins, even in the automotive industry. Thus, mental barriers must be overcome.

Given the enormous relevance of software, it is not surprising that companies like Bosch employ more software developers in India than SAP. Siemens is moving in a similar direction with its 17,000 software developers (and rising), as is VW. The company from Wolfsburg now has over 11,000 software developers on its payroll. The trend toward software and data and away from hardware is unmistakable. However, such transformations carry enormous risks, such as missed deadlines or unmet quality standards.

The realization that the correct handling of data (and the "right" data) is crucial for companies goes back some time. One origin of this is KAIZEN—that is, continuous improvement, see [15].

4.1.3 Economic Relevance

On the basis of (new) data or finally sensibly utilized data, new products and business models are emerging. More on this, as already mentioned, in Chaps. 6 and 8. Data is increasingly becoming a differentiating factor for companies. This also applies to public administration.

The increasing integration of software into products, already mentioned, is particularly evident in the case of automobiles. Here, significantly more innovations are realized through software (e.g., value-added services) than through hardware. This is also associated with software-based differentiation in the market. This differentiation generally occurs in combination with the use of (customer) data. It is therefore not surprising that, according to the former BMW board member Harald Krüger, a lack of or poor connectivity will be the main reason for not buying a car in a few years, see [40]. This can also be seen in the changing purchasing behavior of Chinese consumers, who are, however, harbingers of a general shift in buying behavior. No longer are engine performance and build quality (keyword: panel gap) the decisive purchasing criteria, but rather the connectivity of the "device" car with one's own smartphone.

Another example is agriculture. By using climate data, soil light exposure, soil moisture, soil and surface temperature, supplemented by satellite images, US farmers were able to reduce water consumption by around 35%, see [19]. Such approaches are increasingly being adopted worldwide.

Another example comes from the financial sector. Here, so-called robo-advisors provide digital asset management (i.e., via software programs), especially for clients with a four- to five-figure budget, see [41]. The market volume of such solutions in Germany amounted to around €360 million in 2015. By 2019, it had already reached €7.5 billion, see [27]. As early as 2017, around 30% of all banking transactions were conducted directly over the Internet and 44% of all current accounts, see [37]. In 2021, the number of online banking users in Germany rose to almost 47 million, see [46]. This trend was further accelerated by the coronavirus pandemic.

However, the downsides of software must not be ignored—cybercrime, i.e., attacks from the internet (and increasingly from within the organization). More on this in the next chapter. The economic damage from an attack can reach company-threatening proportions. According to the industry association BITKOM, cybercrime caused damages of €110 billion in Germany in 2017, see [2]. By 2021, this figure had already risen to over €203 billion, see [47]. The issue does not spare even the giants of the industry, including the world's most valuable company, Saudi Aramco, estimated at $2 trillion. In 2012, it fell victim to a major hacker attack, see [33]. The world's largest container shipping company, Maersk, was also hit by a hacker attack in mid-2017. The damage is estimated at around $200–300 million, see [35]. These examples are part of a long list of companies from all sectors worldwide.

4.2 Big Data and Data Lake

What cannot be measured cannot be managed. The large volumes of data already mentioned must be stored and processed both effectively and efficiently in order to ultimately generate added value, as has been discussed several times. Various terms are used for this storage, such as data warehouse, big data, and data lake. Therefore, the first step is to clarify these terms.

Due to the steadily increasing demands in companies for more flexibility and performance, combined with rapidly growing heterogeneity of data and their sources, established technologies such as database systems and data warehouse solutions are often reaching their limits today. The already mentioned huge heterogeneous data volumes (i.e., different data formats) can only be managed using specialized software. In a data warehouse, structured data from upstream systems or an ERP system is typically stored. Structured data has a standardized format for storage (e.g., first and last name or the torque during a screwing process), in contrast to unstructured data. More on this shortly and why it is so important.

With big data, employees can process significantly more data and use it for their decision-making, as they learn much more about their business and company. This knowledge enables them to make much better decisions and, at the end of the day, achieve better economic results, see [23]. Recognizing patterns, i.e., relationships between data, is crucial. The use of big data systems spans all industries. This broad application is also due to the fact that the focus is less on searching for cause-and-effect relationships and more on identifying statistical correlations. The technical component only comes into play in a second wave of optimization. However, the definition of the term big data is continuously evolving.

Another approach to handling data is the data lake. Unfortunately, this term is also not uniformly defined, see [48]. A data lake is a storage system for data analysis. This can be in the cloud or on-premises. All cloud providers offer corresponding solutions. This view is also shared by Gartner. According to Gartner, it is a concept of storage instances for a wide variety of data formats in their original format. The concepts date back to 2010. In a data lake, data from various sources (social media, databases, machines, and equipment) can be stored and saved. Thus, it involves various types of data (referred to as structured and unstructured data), see [49]. In order to quickly access and utilize this data, additional information (metadata) is required.

Big data thus represents the underlying core technology for building a data lake [50]. Therefore, a data lake often includes big data technologies as well as other technologies for handling large and variously structured data [50].

It should come as no surprise that companies need both a business warehouse and a data lake, regardless of whether these are company-wide or specific to a business unit. The use depends on the specific application case (use case).

4.2.1 Technical Description

Since a data lake is generally based on big data technologies, for simplicity, both terms are used together and synonymously. Currently, “data lake” is the term used much more frequently. It is, so to speak, the “new” term.

It should therefore come as no surprise that in the literature, big data and data lake are often equated. But first, let’s look at the term big data. As already mentioned, this refers to software for processing enormous volumes of data in a short period of time, in order to ultimately gain insights from data. These insights are the prerequisite for business decisions and thus the basis for generating competitive advantages. A definition of big data is graphically depicted in Fig. 4.2. In short, it is about the rapid processing

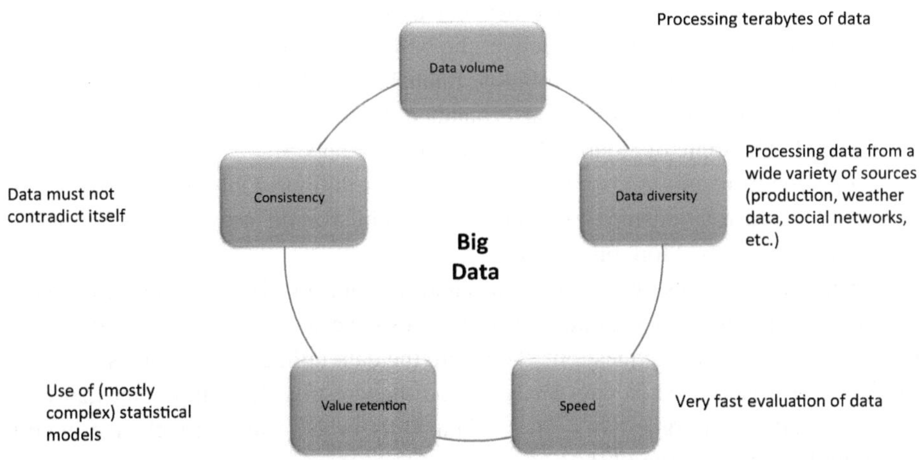

Fig. 4.2 Definition of Big Data

of enormous amounts of data of any kind, i.e., so-called structured data (as stored in traditional databases) but also unstructured data, which may arise, for example, when searching the internet or in production. In this context, the 4 V's are often used: Volume (amount of data), Velocity (speed of data generation and processing), Variety (range of data types and sources), and Validity (value and data quality). Additional V's are sometimes added, such as the correctness of the data [71]. The added value, which has already been mentioned several times, can be generated using classical analytics or with the help of artificial intelligence. More on this shortly.

When it comes to big data, it is important to note that it is not a single technology, but rather a bundle of technologies and their interplay. The analytical processing of the aforementioned different types of data is the actual core of big data systems. There are specialized solutions for generating/extracting data, and analysis is usually performed using statistical models or approaches from the field of machine learning (see more on this shortly). The specific use case is crucial when selecting the technology. Frequently used application scenarios include maintenance (predictive maintenance), quality (predictive quality), or data mining.

A key element of big data is fast response times. For this reason, so-called in-memory solutions (i.e., solutions in which a large portion of the data is kept in main memory) are often used. Such solutions have been advanced, among others, by SAP (S/4 HANA). In these cases, not only the processing but often also the visualization of results takes place in main memory, rather than primarily on the hard drive. This often involves processing very large volumes of data, for example in the terabyte range or larger, such as the previously mentioned exabytes.

To make it easier for users to get started, such solutions are offered in the cloud, i.e., the application is no longer located in the company's own data center (so-called on-premise solution). However, the data usually still remains local to the respective company. It is transferred to the cloud solely for analysis. Complete storage and processing in the cloud are also possible. In this case, corresponding "preconfigured" solutions can be used. While these do not always deliver the best possible results, they do provide very rapid initial added value with relatively low financial and time investment. The operation of cloud-based applications is then also handled by the provider and not by the company's own IT. The topic of cloud will be revisited in Chap. 5.

Visualization of results is also important for such solutions. Only through this can relevant insights and relationships be gained. This point should not be underestimated in corresponding projects. More on this later.

When introducing these new approaches and decision-making methods, the first step is to ask about the data sources and what analyses have been performed on the data, in order to address the crucial question: "What are the data telling you?" This is also linked to the level of trust placed in data analyses and their results. It is important that executives and management argue based on the data. Only in this way can a change in corporate culture be achieved, see [23].

Of course, subject matter experts will not become obsolete now or in the future due to big data and data lake systems. They must support data analysis with their expertise in order to draw the right conclusions from the data analyses.

However, big data is not a sure-fire success. As a prerequisite, consistent, i.e., comparable data with high data quality must be made available, which is not always easy in operational practice. The business processes analyzed using big data approaches must run stably (which is often precisely one of the problems in companies). There are also systems that can handle fuzzy data. However, this does not make things (i.e., implementation) any easier. Fuzzy data are usually a sign of unstable processes. In practice, these two points (data quality and process stability) represent the biggest hurdles to using the technology. However, it is also evident that companies that use big data systems have achieved a corresponding level of process maturity. Thus, technologies are only one part of success.

4.2.2 Consequences for Management

One point at the outset: your previous investments in database-based solutions were not in vain. Rather, they provide a good starting point. However, such solutions are by no means sufficient to maintain or even expand your own competitiveness. They only allow for limited intelligent use of all available, as well as new, data.

It should not be overlooked that the introduction of Big Data systems within your own company certainly presents a technical challenge that should not be underestimated. However, this is only the well-known "tip of the iceberg." Alternatively, such systems can also be used in the cloud with the respective cloud providers. This also applies to a Data Lake, which can be set up in a variety of technical forms. We will discuss the topic of cloud in detail in the next chapter. Generally, implementation in the cloud is preferable, as both the financial and technical effort required for in-house implementation are considerable, as already mentioned.

The much more critical element lies below the surface, in the form of management. Big Data and Data Lake projects represent a management revolution. They change the way decisions are made and should be made, and who has the authority in this process. Dominic Barton (Managing Director at McKinsey) noted this some time ago. Currently, many decision-makers unfortunately still rely on their experience and intuition, or that of their top advisors. A shift is needed towards data-driven decision-making and consciously questioning one's own intuition. However, especially in the early stages of such projects, one should not blindly trust the results and data either.

Thus, managers must also learn to handle and understand Big Data and Data Lake systems. Basic statistical knowledge should be expanded accordingly. The key is to ask the right questions and interpret the answers. A Chief Digital Officer (CDO) anchored in the business unit should be present in companies with around 500 employees or more, as a counterpart to the Chief Information Officer (CIO), to support implementation from the business perspective and drive the topic forward. The topic of the CDO will be revisited and explored in detail in Chap. 9.

When setting up Big Data and Data Lake projects, managers should choose an agile approach to ensure rapid implementation, foster learning within the company, and ultimately be able to present quick initial results. This significantly increases the acceptance of such solutions. To accelerate this and keep investments low at the same time, cloud-based solutions from the respective vendors can be used.

However, data alone does not make a company successful. It is much more important to ask the right questions and interpret the analysis results accordingly. Such systems do not replace vision or creative thinking; they support their validation. An important point when introducing Big Data and Data Lake solutions is to use your own creativity in selecting use cases and their solutions. This is how the corresponding and much-desired competitive differentiators can be generated.

Another consequence is that so-called "Data Scientists," i.e., experts who master Big Data systems and Data Lakes, are becoming increasingly important for companies. Close collaboration with the respective business units is required here, as this is where the use cases arise. This sounds simple—in practice, however, this collaboration often proves to be challenging. Clear rules and responsibilities must be defined. The business unit is naturally responsible for generating economic value and for rollout. In general, knowledge in handling data must be increased in both the business and IT departments. Economic value does not arise automatically; rather, it requires close collaboration between IT and the business unit on an equal footing (the latter is extremely important).

4.2.3 Economic Relevance

The central statement is: With Big Data and Data Lakc, products become better and internal processes more efficient. In addition, new business models emerge from data, regardless of the industry. The automotive industry is certainly a pioneer here, but online retail also demonstrates this on a daily basis. This results from better interaction with the customer. By using existing and new data (keyword: social networks or sensor data, i.e., evaluation of heterogeneous data pools), it is possible to recognize customer needs much more precisely and to derive product changes and services. Thus, data generation supports direct customer contact to identify and respond to customer needs. This enables the creation of customer-specific and personalized offers, as well as the collection of information about user behavior and the acceptance of features, such as special equipment in individual vehicles. Overall, sustainable brand loyalty and improved customer loyalty can be achieved, as switching barriers can be specifically built up. By analyzing various data (social networks, customer feedback, news, etc.), market trends, customer preferences, user habits, and customer needs can also be identified early and, if necessary, taken into account in product development. Again, the automotive industry is a key driver. According to consulting firm Bain, about 90 million connected vehicles were sold in Europe between 2015 and 2020. Additional data can be generated via the digital services available through these vehicles.

Of course, a distinction must be made here between B2C and B2B business relationships. Appropriate approaches must be developed for each segment. However, the relevance remains the same.

The areas of application become clear from what has been said so far:

- Corporate management,
- Logistics,
- Finance and controlling,
- Sales and marketing,
- Product development,
- Production,
- Maintenance.

Applications in the field of HR are certainly interesting, and there have already been some initial practical successes in this area, but to what extent it is an effective tool for recruiting new employees remains to be seen. However, the areas listed above are clearly more strategic. Usage is also often found in the area of procurement.

To make Big Data and a Data Lake economically successful, a data-driven integration of the entire IT landscape is required. All levels of the company must be activated to generate data and to create derivations and cross-references. To improve statistical models, a combination with machine learning has recently been used. More on this shortly. As already mentioned, it is also important to build up the necessary knowledge in the respective business units, as this is where the use cases arise—or do not.

In the aviation sector, for example, Korean Air uses the IBM system to reduce the time required for troubleshooting defects. Here, time expenditures were reduced by around 90%. This also increased the punctuality of the aircraft, see [13]. Similar improvements were achieved at the Gallo winery in the USA with irrigation control. Water consumption was reduced by 25% while simultaneously increasing wine quality, see [14]. These examples demonstrate the diverse application possibilities of Big Data solutions—regardless of the manufacturer. These experiences are also consistent with McKinsey's forecasts. The consulting firm estimates savings in maintenance costs in production through the use of predictive analytics approaches at 10–40%, while simultaneously reducing production downtime by around 50%, see [24]. Over the years, these forecasts have been substantiated by concrete examples, see [70]. However, it must be mentioned that such studies can only provide an indication of possible savings in one's own company. In practice, a first top-down approach should be taken based on such studies. This allows for the rapid identification of possible concrete savings. In the next step, these savings are verified using a bottom-up approach. Unfortunately, this approach is much more complex.

4.2.4 Risks

Storing and processing personal data present a particular challenge in Big Data projects. It is important to involve the works council at an early stage. Legal aspects must be taken into account right from the start of the project. These can lead to significant project delays or increased costs and complexity.

When implementing software-based solutions in general, and Big Data and Data Lake solutions in particular, it is important to ensure that the business model behind the solution is not easily copied or is difficult to replicate. Software as such can, of course, be copied very easily.

In addition, there is always a certain risk of failure and a risk of misinterpretation, i.e., the analyses are based on incorrect assumptions, data, or mathematical models. Before going live with such solutions, they should therefore be thoroughly verified and tested from a technical perspective. Here, the technical solutions form a basis for later evaluations and analyses. More on this, as already mentioned several times, shortly.

The banking sector is actually predestined for Big Data systems and will certainly radically change the retail banking business, see [33, 41]. On the other hand, the unchecked use of data also leads to economic risks—such as in stock trading.

In general, it can be observed that entire industries are facing a complete transformation, or even their demise, whether the industries are aware of it or not. Overall, Big Data and Data Lake will lead to significant disruption in employment structures. It is therefore all the more important to consistently further qualify your own employees. More on this in Chap. 9. However, this disruption is only possible in combination with other technologies, such as Artificial Intelligence/Machine Learning. Thus, the already frequently mentioned (correct) networking of individual technologies is of central importance.

4.3 Digital Twin

The concepts for a digital twin are not truly new. They date back to 2002, when a digital representation of real objects in product development was first postulated, see [51]. This brings us to its definition. A digital twin is the representation of a physical object in the virtual world. Thus, a digital twin does not consist solely of data, but also of applications. Examples include vehicles, production lines, or, more abstractly, entire supply chains. Ideally, both worlds (real and digital) interact with each other. In 2018, Gartner identified the digital twin as one of the 10 most important technology trends. However, a prerequisite for the successful use of digital twins is the data that has already been introduced. This is absolutely essential. The core of a digital twin is data.

Since its beginnings over 20 years ago, the topic has evolved: Currently, a distinction is essentially made between three types of digital twins:

- Product design (product twin)
- Production planning and validation—that is, the production equipment (production twin)
- Production—that is, the representation of each individual manufactured product (performance twin).

If one considers the lifecycle of a product, the execution twin is still missing from this perspective, i.e., when the product is in use, such as a car, a watch, or a smartphone. In total, the entire structure of a company can be mapped using corresponding digital twins.

However, one should not fall into the misconception that there is only one twin for each category as listed above. This may still apply to product design and its planning, but the situation is quite different in production. Here, a multitude of twins exist, as every car or smartphone that is produced receives a digital representation.

The situation becomes even more complex because, for all three digital twins listed above, they are hierarchically refined to adequately represent the real object. However, the goal is not necessarily a 1:1 mapping of the real object in the virtual world. The aim is to represent the characteristic properties and features in the digital world. Figure 4.3 Digital Twin provides insight into the hierarchical structure. Thus, the application area of digital twins is extremely multifaceted. Depending on the area of application and the use cases considered, their implementation can become extremely complex. Therefore, here too, the complexity of applications should be increased gradually over time. At the end

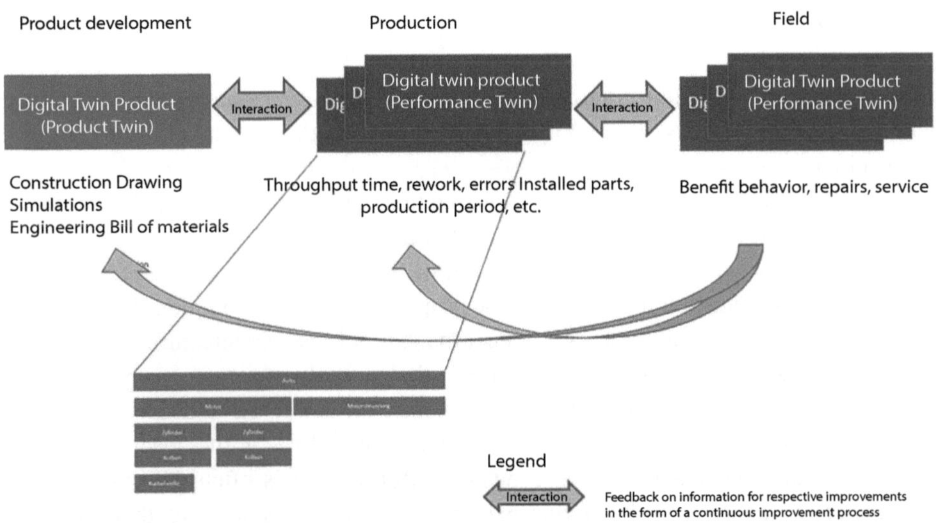

Fig. 4.3 Variants of the Digital Twin

of the day, it is about storing data through which a digital twin is represented. This storage takes place in appropriate databases. In the production environment, the corresponding twin is built by (hopefully) automated data collection during the production process. Depending on the product, several thousand or tens of thousands of data records are generated per product.

4.3.1 Economic Relevance

Due to the broad range of applications, it is difficult to make general statements regarding economic savings potential. Therefore, the demonstration of economic added value is based on specific scenarios.

According to the above, the main areas of application for a digital twin are in product development, production planning, and production. A digital twin is the starting point for further applications, such as predictive maintenance or condition-based maintenance. In addition, capital efficiency is increased through optimal use of assets (such as machines, trains, elevators, or offshore wind turbines) and personnel. This again demonstrates the wide range of applications for the technology.

In the area of production, throughput times can be improved by around 30%, see [52]. These statements are generalized observations. In one's own company, the initial situation must of course be taken into account, and thus the added value must be determined individually for each company. In production planning, the use of digital factory models, which lead to the creation of a corresponding digital twin, can also reduce efforts by about 30% while simultaneously reducing project risks.

4.3.2 Risks

In the area of human-machine communication, as well as in data storage, there is a threat scenario from cyber-attacks. This applies regardless of whether the data is stored locally or in the cloud.

When extending the digital twin to organizations or people (human digital twin), the GDPR sets boundaries. However, this is also a very specific variant. Depending on the use case, a more or less complete mapping of the real world in the digital world is required. The more detailed and granular the mapping, the more complex the undertaking and the higher the requirements for governance and data architecture. However, it must be mentioned that the added value of a digital twin does not necessarily increase with its level of detail. The higher the requirements, the lower the added value. Therefore, the focus when determining the level of detail should be on the specific use case and not on what is technically feasible, which unfortunately is often the case in business practice. In general, digital twins exhibit a complexity similar to their physical counterparts.

The full potential of a digital twin is not realized in its isolated form within a business unit (such as production or product development), but through the creation of networked ecosystems. This allows company boundaries to be overcome and new ecosystems to be built. This is especially true for OEMs (Original Equipment Manufacturers). These exist in all industries—not just in the automotive sector. However, this requires cross-company win–win scenarios, which need to be developed.

In practical implementation, a variety of solution providers compete for customers' favor. For the areas of product development and production, many of the solutions offered are at a usable level of maturity. Naturally, there is some catching up to do with cross-company solutions. These should only be tackled in a second or third step. In the first step, the already mentioned areas of product development and production should be the focus of implementation. It is important to ensure that data ownership (i.e., who owns the data) is clearly defined. The data should belong to the customer and not to the product manufacturer. Therefore, open systems that can be integrated into one's own system landscape should be clearly preferred. Unfortunately, some product manufacturers see this differently (which is understandable from their perspective, but is not really a future-proof approach). Regardless of the risks, the implementation of individual digital twins should be at the top of the priority list in the digital transformation of companies—regardless of the level of detail. As with many topics, the most important thing is to get started. Gaining experience within one's own company, learning from possible mistakes, and strengthening one's own level of innovation should be the recommended approach. By taking this step-by-step approach (i.e., starting "small"), the risk of technical failure can also be reduced. Also, the individual twins do not necessarily have to communicate with each other at the beginning.

4.4 Artificial Intelligence/Machine Learning

Artificial Intelligence (AI for short) has been familiar to everyone at least since Apple's Siri assistant system. Such "helpers" are now encountered not only on smartphones, but also, for example, in cars or smart TVs. Autonomous driving has been considered a game changer for many years. However, the term "artificial intelligence" is somewhat misleading in this context. Therefore, practitioners (including the former CIO of VW—Mr. Hofmann) prefer to use the term machine learning. The goal is not to replicate the human brain, but rather to enable machines to build their own competence, see [53]. Thus, it is about pattern recognition in existing datasets, predicting events, and making optimizations. The aim is to identify causal relationships in data. The systems are therefore intended to solve problems more or less independently. The degree of "intelligence" is directly linked to the level of autonomy, the complexity of the problem, and the efficiency of the solution method. There is therefore not "the" intelligence, but rather degrees of intelligence, see [22]. Machine learning is one of the most important universal technologies in the context of Industry 4.0. All of these current systems share the characteristic that they are not capable of reflecting on their own actions. This is reserved for the next stage of evolution, see [54]. There is ongoing debate about when such systems will exist. The same applies to when such systems will reach or even surpass the capacity of the human brain. It will likely be several more decades before this happens.

Experience from business practice shows that the use of the term "artificial intelligence" unnecessarily raises concerns within companies. Therefore, the term "machine learning" should be used instead—also because it describes current activities much more precisely, as already mentioned.

The capabilities of such AI systems have been evident not only since ChatGPT. They currently surpass humans in areas such as handwriting recognition, speech recognition, image recognition, and reading comprehension, and of course in processing large volumes of data and the patterns contained therein. However, outside of these areas, the performance of AI systems is very inconsistent, see [36].

4.4.1 Technical Description

As an academic research field, AI has existed since the 1950s. It is therefore not a new approach. A wide variety of technological approaches are hidden under the umbrella of machine learning. Only the availability of sufficient computing power and access to correspondingly large datasets has enabled the technology to break through in recent years.

Mathematical models or neural networks are often used. Neural networks are, broadly speaking, inspired by the human brain (to say they are modeled after it would be an overstatement). The focus is on empirical, data-driven approaches. As a result, training them requires correspondingly large datasets. The biological metaphor has increasingly faded into the background. Multidimensional networks are often used. Both structured and unstructured data can be processed. The more neurons are used, the more data is required. As a rule of thumb, about 10 data points are needed per neuron, see [32]. For sufficiently large networks, this results in a significant demand for data. Often, a seven-digit number of datasets or more is required to obtain meaningful results and thus trained models. A varying number of hierarchical layers is used. The first layer serves for data input and the last for outputting results. The more additional layers there are, the more flexible and accurate the system can act.

An important example here is IBM Watson. Such systems are often referred to as cognitive systems. Only in this way can the individual neurons deliver meaningful results. This also creates a bridge to Big Data. These systems provide the large datasets required. Watson became famous for winning competitions against grandmasters in chess or Go. In addition, large language models (LLMs) such as ChatGPT have gained importance. These can be used to implement chatbots, make extensive documents accessible to users via questions, and also to capture lessons learned. All these documents are often in the form of Excel sheets and thus become increasingly difficult to use as their number grows.

The second approach is based on so-called "deep learning" (self-learning algorithms), i.e., systems that learn from their own behavior and mistakes. This is very closely modeled on human learning by imitation. The focus here is on feature extraction, pattern recognition, and feature formation. The behavior is therefore no longer specified by programmers or engineers, but is the result of learning processes. This represents a paradigm shift. The (learning) quality and thus the results and decisions of such systems depend very much on the training examples and thus the training and test data. Prejudices or value judgments that have been implicitly or explicitly incorporated into the training phase also affect the result. In a production plant that optimizes itself, in the final analysis, rejects would also have to be allowed, which can lead to corresponding problems, see [32]. Thus, the statements of such approaches depend less on the algorithm or mathematical model, but rather on the chosen training data, which makes their use in an operational environment more difficult.

Both approaches unfortunately share the characteristic that, especially in the production environment, there are still no systems in productive use in the form of self-control. However, the approaches are already suitable today for supporting decision-making, see [32]. The company DeepMind, acquired by Google, has developed a learning algorithm that taught itself to play 49 different Atari video games, see [16]. There are various funding initiatives in the industrial context to establish self-control in operational production. In general, AI-based approaches are currently widely used, including in production. One application is quality control in press shops or for paint defects using camera-based approaches. In another application, AI helps with risk analysis of supply chains, see [55].

Another example is so-called "fuzzy" systems, i.e., systems with fuzzy logic. They can handle uncertainty factors, i.e., imprecise statements. One can vividly imagine this as calculation with probabilities. Of course, temporal changes must also be taken into account.

The most commonly used approaches in production practice are predictive maintenance and predictive quality. However, the importance of predictive maintenance is often overestimated. Here, effort and benefit must be weighed very carefully. A wear-based approach also delivers very good results with lower complexity. In addition, camera-based approaches are frequently used. This again creates a bridge to Big Data. For the pattern recognition required here, proven algorithms exist, but they must be adapted to specific needs. This is especially true for plant engineering, the rail industry, or the aerospace industry, as unique machines are often used in these sectors. Most models for determining remaining service life or predicting product quality are based on statistical analyses, such as the failure history of individual components. The accuracy increases with the effort invested in developing the respective models. If only a few sensors are available, this approach is certainly the most suitable. For so-called data-driven approaches, real-time availability of data is an important prerequisite. It is then possible to make forecasts for a specific time horizon based on statistical models.

Another very popular example of AI systems is Robotic Process Automation (RPA). This involves the automation of business processes using so-called software bots. These systems can also use self-learning approaches to increase efficiency.

4.4.2 Consequences for Management

Among employees, the term AI often evokes great skepticism, if not outright rejection. This fear stems from the potential loss of one's own job if the "machine colleague" takes over their tasks. In general, however, such systems deliver their greatest benefit and added value when interaction between humans and AI is achieved. Regardless of the maturity of machine learning, these systems can support people in their decision-making. Thus, such systems generally support human activities. Complete automation of individual business processes therefore only occurs in specific areas. These systems demonstrate their greatest strengths particularly in processing large volumes of data according to certain patterns and in establishing statistical correlations.

These systems cannot replace inspiration, innovation, emotional intelligence, or social intelligence. Moreover, they place high demands on system, process, and data security, see [16, 42]. The relevance of this topic is also reflected in the fact that companies such as BMW, Bosch, and Daimler are increasingly seeking and hiring AI experts, see [34]. The maturity level of such systems has already been discussed. However, such solutions should not obscure the fact that individual customization and the necessary data volumes are required. Equipment in production is subject to individual stresses and requirements, so generic solutions should be approached with a certain degree of skepticism. Therefore, companies should increasingly invest in data quality and such solutions. However, systems alone are by no means sufficient. It is essential to inspire and qualify one's own employees for this topic. More on this later, as it is a very complex and time- and resource-intensive undertaking.

AI or machine learning-based approaches can be used in virtually all areas of a company. Therefore, it is important to build up the necessary competencies within the company itself. This applies both to IT and to the respective specialist departments. Incidentally, according to a survey, "robo-bosses" are very popular (two-thirds of respondents stated this), as they are unbiased, see [51]. However, a major disadvantage is that they do not understand emotions.

4.4.3 Economic Relevance

The areas of application for machine-based systems are very diverse. Versicherungskammer Bayern, for example, uses Watson to automate and optimize its complaint management. Over 7 million inquiries and complaints from a wide variety of media (email, letter, call center, etc.) must be processed and analyzed quickly. The voice recognition systems Sire and Alexa accompany us in many aspects of daily life. Both are very prominent representatives of these technologies. Autonomous driving in cars is another example, regardless of when it will be fully implemented on German roads.

BMW aims to significantly enhance the personalization of its vehicles and develop intuitive driver assistance systems using Watson, see [26]. Such solutions are also used by the automaker in production. According to BITKOM, over 50% of surveyed companies already use such solutions, see [56]. The Chinese start-up NIO is moving in a similar direction. Here too, deep learning systems are used in the area of driver assistance systems.

Machine learning is reaching an ever greater level of operational maturity. Therefore, it is not surprising that the digital association Bitkom projected high growth rates in this area, reaching around €4.3 billion in 2017, see [32]. According to Spiegel, global AI revenues in 2023 amounted to around USD 500 billion, see [69]. According to a Microsoft study from 2019, this sector is growing at double-digit rates, see [57]. The consulting firm McKinsey also sees significant potential, especially in the areas of predictive maintenance and quality. The consultants estimate that the consistent use of intelligent robots and self-learning systems could increase gross domestic product by about 4% or €160 billion. This corresponds to annual growth of 0.25% or €10 billion, see [18]. Consequently, the consultants also predict that AI could become a growth engine for German industry. It is not surprising that the greatest potential is seen in the automotive industry, aviation, mechanical engineering, and the semiconductor industry.

All studies agree that AI-based approaches will contribute significantly to productivity increases in companies. Therefore, managers are well advised to look for application scenarios in the individual departments. This could determine the future competitiveness of their own company. Again, it is advisable to start with use cases of manageable complexity and use these to build up expertise within the company. According to a McKinsey study, generative AI (i.e., LLM) has an economic potential of around USD 2.6 to 4.4 trillion, see [68].

4.4.4 Risks

The use of deep learning systems in particular carries significant risks due to the aforementioned error rates. This results from the extensive and complex testing required. For example, the Australian tax authority introduced such a system called "Centrelink" to review payments. At the end of 2016, over 160,000 reminders were automatically sent by the system, unfortunately with a high error rate. Only 20% of the claims were actually justified. The USA had a similar experience with software used to assess the risk of recidivism among offenders. Here, a direct relationship was drawn between skin color and risk of reoffending, see [38].

These examples clearly demonstrate the technical dangers of such systems. Great care is required in training and therefore also in testing. Fully automated decisions and controls using deep learning systems for complex issues without human interaction should, at least for now, be viewed critically.

Therefore, an explanatory model is needed for AI-based solutions. The acceptance of such solutions in everyday business life depends on this. The potential is considerable, especially in the medical field (e.g., care), but also in the transport sector, see [67]. This also shows that there is increasing societal acceptance.

The greatest risk, however, lies in doing nothing or being discouraged by initial, less successful projects. It should not be overlooked, however, that AI-based systems are increasingly being used for military purposes. Thus, the military, especially in major powers but also in countries like Israel, has recognized the potential of this technology, see [21]. To what extent such developments actually pose a threat remains to be seen. However, there will be an "AI arms race" in the military sector.

4.5 Cyber Physical Systems

"Cyber-Physical Systems (CPS) are characterized by the linking of real (physical) objects and processes with information-processing (virtual) objects and processes via open, partially global, and always interconnected information networks," see [3]. In short, CPS are products that have emerged through an intelligent combination of hardware and software, with one special feature—they can communicate. This communication takes place either via the company's own network or, for example, with the cloud (of the manufacturer or their own). Much of the above definition is not truly new. The novel aspect is the networking. This enables an almost unlimited connection with other process participants and allows (at least indirectly) access to the knowledge and experience of other companies. An example is the Augsburg-based robot manufacturer Kuka, which offers its customers cloud-based solutions as functional extensions for its sensitive robots. This makes adaptive and self-controlling, self-configuring, or self-optimizing systems and products possible. If used within the company itself, this opens up the possibility of building a decentralized and partially self-controlled infrastructure instead of a central and hierarchical communication (i.e., from ERP to MES and from there to the machine and finally to the product).

From what has been said, it becomes clear that, strictly speaking, the sensitive robots discussed in this chapter (see Sect. 4.6) and the autonomous transport units described in the next chapter are concrete representatives of this new "species." Other examples include "intelligent" machines and systems that, for instance, independently organize necessary maintenance and repair activities. "Intelligence" here is achieved through the AI or machine learning already mentioned in the previous section.

4.5.1 Technical Description

Integrating CPS into the Internet or using cloud services requires an appropriate infrastructure (regarding cloud, see Chap. 5). More advanced services or even the entire logic are then located in the cloud. The schematic representation is shown in Fig. 4.3, see [28].

Programming often relies on specific embedded software. Communication between individual CPS and/or the environment takes place via machine-to-machine communication (M2M). Proven standards exist for this. The high complexity in developing CPS solutions is due to the large number of diverse technologies and system components that must be mastered. Yet, it is precisely this efficient interplay of hardware and software components that distinguishes such solutions. IT security is particularly important. This must be taken into account already during product development. If the products are production machines or systems, a so-called OPC-UA (Unified Architecture) server must be provided for communication. This allows easy integration into the entire Industry 4.0 production environment; see also vertical integration in the next chapter.

A central aspect is smart sensors and actuators. They are the "eyes and ears" of CPS. Through them, active interaction with the real world takes place. Such sensors are called "smart" because they possess a certain degree of intelligence. They are able to actively perceive their environment and, based on this, initiate appropriate actions. Popular examples are sensors in the household sector, such as intelligent door openers. Here, a smart sensor not only records individual measured values but also processes them. For this purpose, such sensors have corresponding microprocessors in combination with standardized interfaces to higher-level systems. In production, such sensors can detect the individual occupancy of a workpiece carrier and thus determine which slots are available [58]. This is further supported by the fact that such sensors are subject to a continuous decline in price.

4.5.2 Consequences for Management

CPS enable the creation of entirely new products. The software component can be as large as desired. It is also possible to "refine" existing products with sensors and software. Examples of CPS include wind turbines. Control can be carried out via the cloud, see [29]. Sensors can handle the connection to the cloud.

Cars and trucks are also increasingly becoming CPS or mobile computers. For Apple, cars are the largest devices. A similar "fate" awaits airplanes. These too are being equipped with more and more sensors and are delivering ever more data. Once again, the revolutionary aspect is not the individual technology, but the possibility of combination or networking, such as the generation and thus processing of data using big data systems. Aircraft turbines and power plant turbines are also designed and used as CPS. Car roofs can also be made "smart" using sensors, thus becoming an important element on the path to autonomous driving. Companies would therefore be well advised to pay increased attention to this field.

Due to the computing power required, such approaches would not have been possible a few years ago. The increased use of sensors and their communication capabilities makes it possible to offer customers not just the respective machine or system, but their original need in the form of machine availability or throughput. Companies are therefore increasingly selling services.

4.5.3 Economic Relevance

The use of CPS offers several advantages for in-house production:

- Optimization of maintenance (predictive maintenance) including self-diagnosis, and thus machine utilization and productivity increase while simultaneously reducing costs,
- more flexible response to production and product disruptions, such as offshore wind turbines or aircraft turbines,
- better integration of product development and suppliers into production (horizontal and vertical integration, see Chap. 5)
- construction of more complex subsystems consisting of hardware and software, thereby improving structuring,
- more intelligent process interlocking, resulting in improved quality,
- greater flexibility in production,
- simple process mapping through direct communication between individual machines.

As is often the case, it is the combination with other technologies that makes the difference.

The relevance in the product area has already been pointed out several times.

4.5.4 Risks

The risks associated with a decentralized and CPS-based approach for in-house production are more or less obvious, see among others [9].

- IT/OT security and functional safety of production systems,
- Planning capability, reliability, and availability,
- Development, testing, and commissioning,
- Resource and cost-optimized control,
- Traceability of production results,
- Lack of know-how in dealing with CPS.

Of course, software-based functionalities can be copied much more easily than hardware-based ones. Effective protection against plagiarism is therefore essential. On the other hand, it is important to realistically validate the potential and initiate appropriate product developments.

4.6 Sensitive Robots

The use of novel robots goes far beyond that of classic industrial robots. Autonomous vehicles, robotic ships, robotic systems for space exploration, swarm-based robots—mostly in the form of mini-robots—service robots in healthcare, vacuum cleaner robots, or robots for the tedious task of mowing the lawn are just a few of the many examples of application areas. This is also associated with the problem of a clear and precise definition of the term robot.

Sensitive robots, also known as cobots (for collaborative robots), are increasingly taking over tasks, including in the service sector. BMW uses humanoid robots in production at its plant in Spartanburg, USA, see [66]. Toyota has developed the care robot Robina. In addition, there is Humanoid as a counterpart in the household sector. This also reflects the fact that Japan is currently number one in robotics research. The demand for such "helpers" can also be seen in the numbers. In the EU, the proportion of residents over the age of 65 will rise from 17% to 30%. This demographic shift is also reflected in other countries such as the USA, Japan, or China. There is therefore a great need for support to keep the costs of an aging population under control. The topic of demographic change has already been addressed in Chap. 2.

In the field of robotics, Kuka has defined four stages of robot development (analogous to the four stages of the industrial revolution). In parallel, however, there are also three stages of automation development, see Fig. 4.4. The presentation is based on Davenport and Kirby, see [6].

Regardless of the specific form, robots on the one hand substitute humans through increased automation. On the other hand, robots are increasingly becoming colleagues for humans. Here, too, the issue of acceptance must be considered, and human-machine interaction (HMI for Human Machine Interface) must be seen as a critical success factor. Cobots make an active contribution to overcoming the current shortage of skilled workers. They compensate for demographic change in companies by at least partially taking over tasks from their human colleagues. Thus, in addition to the cost argument, another element in the form of a strategic approach must be taken into account.

4.6.1 Technical Description

The topic of robotics is also increasingly merging with other technologies listed here. Robots are becoming ever more "intelligent" and autonomous. Thus, there are strong

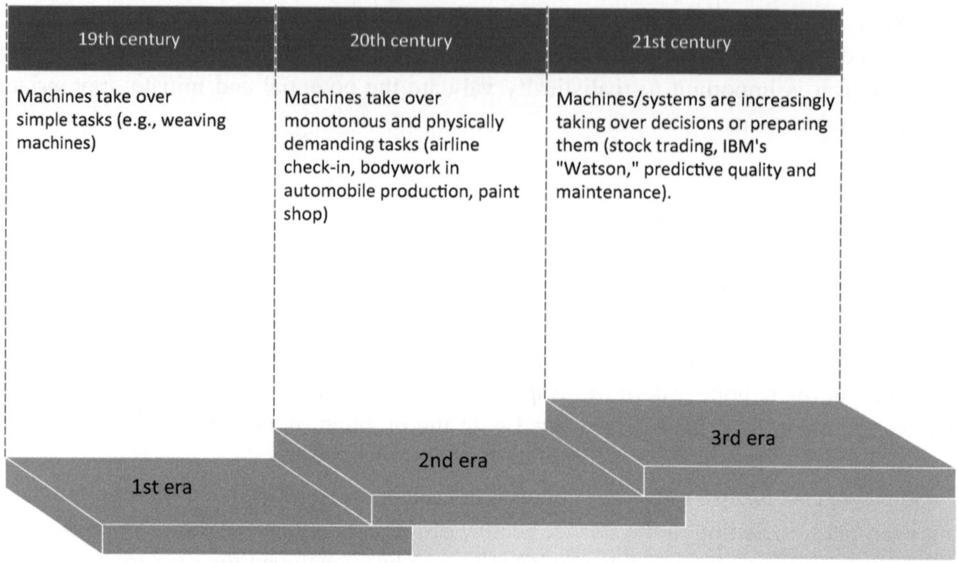

Fig. 4.4 The three ages of automation

overlaps with machine learning and, consequently, with big data. This enables the approach of automation or robotization of production to unfold its full potential.

Due to the aforementioned wide range of possible applications, a general description in brief is difficult. Therefore, the focus here is on sensitive industrial robots (cobots).

Until a few years ago, the use of industrial robots was limited due to safety regulations and their restricted flexibility. If a robot was to take on other tasks, corresponding programming was required. The ever-increasing demands on production (e.g., adaptability, batch size 1) are pushing this concept to its limits. The development of so-called sensitive robots—systems with sensitive joint torque sensors, lightweight construction, force sensors for human-robot interaction, and sensor-based spatial monitoring—has provided solutions to the challenges at hand. Examples of sensitive robots and their applications are shown in Fig. 4.5.

For example, Kuka robots are equipped with a soft and cushioned foam cover, tactile and capacitive proximity sensors. When a person approaches, the robot's movement is automatically slowed down, and in the event of contact, it is immediately stopped. In addition, cameras ensure that the robot's gripping of parts can be monitored. It can thus "regrip" and make corrections autonomously.

The advanced and deployed sensor technology of sensitive robots opens up further possibilities for automation in the form of so-called robot farming and matrix production, see Fig. 4.6. As one might expect, robot farming involves the use of several sensitive robots. These novel robot concepts can be very well combined with the concept of flexible production islands from the world of lean production. Depending on the required

Fig. 4.5 Examples of sensitive robots

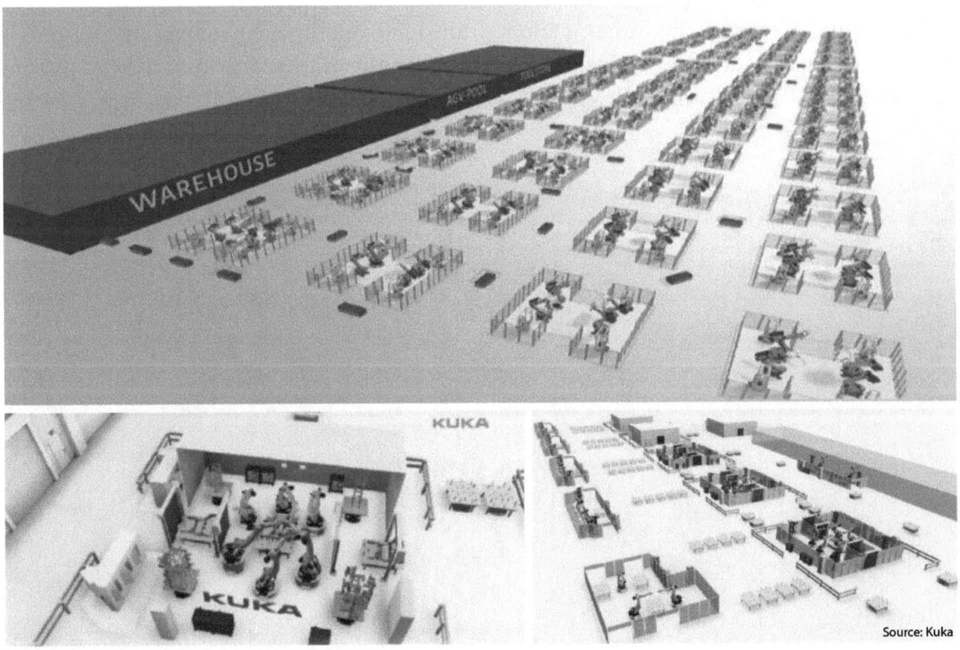

Fig. 4.6 Robot farming and matrix production

quantity and production volume, the employee is supported by one or more robots in these concepts and deploys them flexibly at various stations. The production cell is thus configurable and scalable.

Matrix production is based on standardized production cells. However, in addition to product-neutral equipment, the sensitive robots are equipped with specific tools. This allows for simple and, in some cases, autonomous reconfiguration of the respective robots and thus the cell.

Cloud robotics accesses large volumes of data. This makes it possible to "learn" from the "experiences" of other (identical) robots in similar environments. As a result, optimizations can be implemented more effectively and failures avoided. Until now, robots were limited to their own data. For the development of cognitive abilities in robots, big data and the cloud represent a quantum leap, see [33].

So-called bionic robots take nature as their model. The aim here is not only to emulate humans (humanoid robots), but also, for example, animals, see Fig. 4.7. Swarm intelligence is also borrowed from the animal world. This enables robots to optimize their deployment autonomously. It may sound like science fiction, but it is already a reality. In Germany, the company Festo, among others, is conducting intensive research in this direction. Such systems have already proven their suitability in laboratory operations, see [20]. Potential application areas for such robots include logistics, agriculture, healthcare, and production.

In combination with the aforementioned artificial intelligence (including the previously mentioned LLMs), significantly improved application scenarios arise. AI essentially forms the "brain" of such systems. The novel senses of these "sensitive" robots

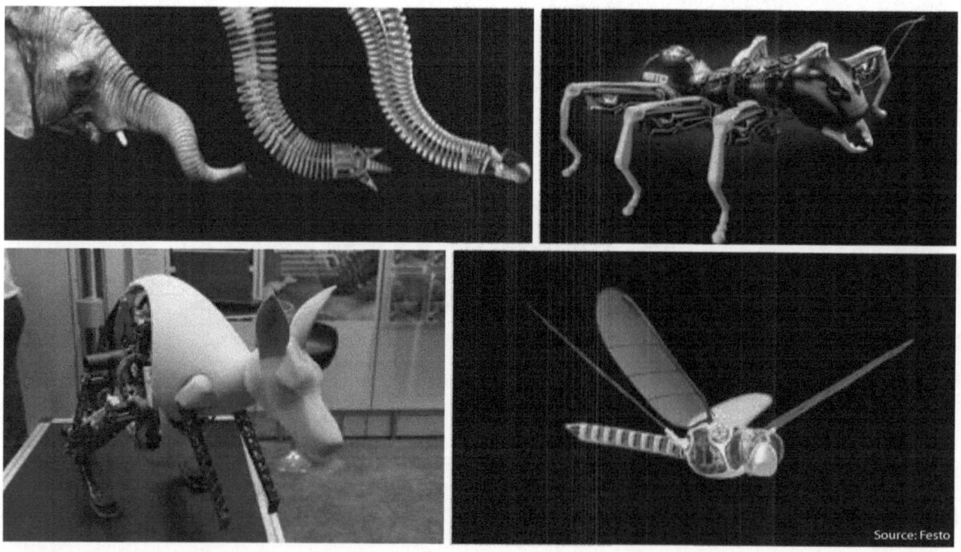

Fig. 4.7 Examples of bionic robots

are provided by the already mentioned smart sensors. Here, too, the already frequently mentioned pattern of combining novel technologies emerges: robotics, AI, and sensor technology. This enables the development of innovative solutions. Incidentally, this also applies to “hearing” and “speaking”—sensors exist for these as well, in combination with appropriate software solutions (not just AI). This results in entirely new possibilities for interacting with robots in a way that is natural for humans.

Robots are also used in medicine. In the USA, the number of medical procedures assisted by robots increases by around 30% per year. It is therefore not surprising that surgical robots are finding more and more customers in this sector. Specifically, around 1300 such robots were sold in 2013, see [33]. In the medical field, robots are capable of acting with great precision and significantly surpass human abilities. In 2019, the revenue from professional robots in this sector exceeded USD 11 billion, see [65].

The German retail sector is also increasingly embracing such robots. The aim here is to make employees’ work easier, counteract the shortage of skilled workers, and, of course, save costs and time, see [12].

4.6.2 Consequences for Management

The advantages of such systems are not limited solely to further automation. Ergonomically unfavorable workstations for employees can also be optimally redesigned. In conjunction with improved human-machine interaction, older employees can continue to be employed in production with high productivity. In this regard, appropriate support from the company’s own works council can certainly be easily achieved. Furthermore, the “mixed calculation” of human labor costs and the costs of deploying robots creates a completely different situation regarding wage and production costs in high-wage countries such as Germany. This makes these locations economically much more attractive again, with increased flexibility and simultaneously reduced response times to market changes. In the case of relocating production, for example to China, logistics costs and response times increase significantly compared to a location in Europe. The use of sensitive robots drastically changes the situation to the disadvantage of locations like China. In addition, this can help reduce dependency on a single production site. However, a general relocation from China back to Europe, and specifically to Germany, unfortunately cannot be expected.

When introducing sensitive robots, for example as part of a pilot project, the mistake should not be made of conducting a strict economic assessment of the isolated pilot. High costs will certainly be incurred here, which would argue against economic deployment. Only the widespread rollout of such concepts leads to economically positive effects.

In general, a payback period of approximately 2–3 years can be assumed in Western European countries. This is a rough guideline, which necessarily depends on the specific application case.

Regardless of the undoubtedly existing economic potential, the works council should be involved in such planning at a very early stage. Understandably, there are significant concerns on the part of employees and their representatives. Job cuts have never been a good marketing argument. Instead, the goal should be to increase sales and thus output from production with the same number of employees. Only in this way can such projects be realized. If there are job cuts, even on a small scale, unnecessary resistance from the works council and a lack of acceptance among employees will be triggered.

In general, such robots represent a good alternative in the face of the current and ever-increasing shortage of skilled workers. However, it must be taken into account that the maintenance effort increases with such complex systems. Additional know-how must therefore be developed. This can be counteracted by optimizing maintenance activities, for example through mobile solutions or predictive maintenance.

In the "Big Five," see [33], that is, Germany, Japan, South Korea, the USA, and China, around 70% of all robots produced worldwide are sold. Most robots are sold in Japan. Interestingly, Russia plays no relevant role in this market, either as a producer or as a market, see [33]. It should also not be overlooked that less developed countries could achieve a leap in innovation through robots. A similar development has already occurred with telephony. Some African countries skipped landline technology and went straight to mobile technology. Thus, these countries could quickly develop into competitors for industrially manufactured goods or become attractive production locations. Robots are also being built in Africa, and at extremely low cost.

When introducing (novel) robots, the respective mentality must of course also be taken into account. In Asia, robots are viewed much more positively than, for example, in Europe, where they are often seen as soulless machines that bring disaster or get out of control.

4.6.3 Economic Relevance

Robots have become indispensable in modern industrial production. The new sensitive generation opens up further automation potential. This ranges from final assembly in the automotive industry, logistics (such as handling and material supply), to service robots, for example in hospitals and for cleaning. In 2016 alone, the American automotive industry installed 17,500 industrial robots, more than ever before. Due to the coronavirus pandemic, sales figures stagnated in 2021. According to Kuka, since 1980, the costs for robots have decreased to 20%, performance has improved by 300%, and required maintenance has improved by 30%, see [1]. With their ever-broadening range of applications, robots are conquering more and more fields. At Amazon, around 1300 robots in logistics centers bring goods to employees. The intelligent robots find the right boxes, load them, and transport them to the packing stations. To this end, Amazon purchased the robot manufacturer Kiva for around $775 million. In addition to the robots already in use, a further 18,000 are planned or being implemented, see [21]. In 2015, robots with a total value of approximately $10.78 billion were in use worldwide. By 2025, this market is expected to rise to nearly $67 billion (including service robots), see [16].

By using such novel approaches, "production elasticity" can be increased by around 30%, enabling the realization of a scalable factory. In addition, there is a significantly improved process reliability with increased automated process documentation, resulting in greater transparency in production. Another advantage of the significantly improved human-robot collaboration is the considerably reduced space requirement compared to isolated human and robot workstations.

In economic assessments, the potential for reducing energy consumption is often not taken into account. With appropriate optimizations, improvements in the double-digit percentage range can be achieved. Furthermore, material consumption, for example in spraying processes, is often significantly lower, which in turn benefits the environment.

China represents the largest growth market for robots. Since 2005, sales there have increased by around 25% annually. China is also developing sensitive robots and is promoting their development until 2025 with around €5.4 billion, see [33].

The prices for selected manufacturers are listed in Fig. 4.8, see [36].

The use of sensitive robots/cobots offers potential for reducing labor costs by at least 30% in high-wage countries like Germany. This results from estimated labor costs for workers of around €50 and for robots of around €5–8, see [39]. This significantly narrows the gap with low-wage countries such as China. This point has already been briefly touched upon in this chapter.

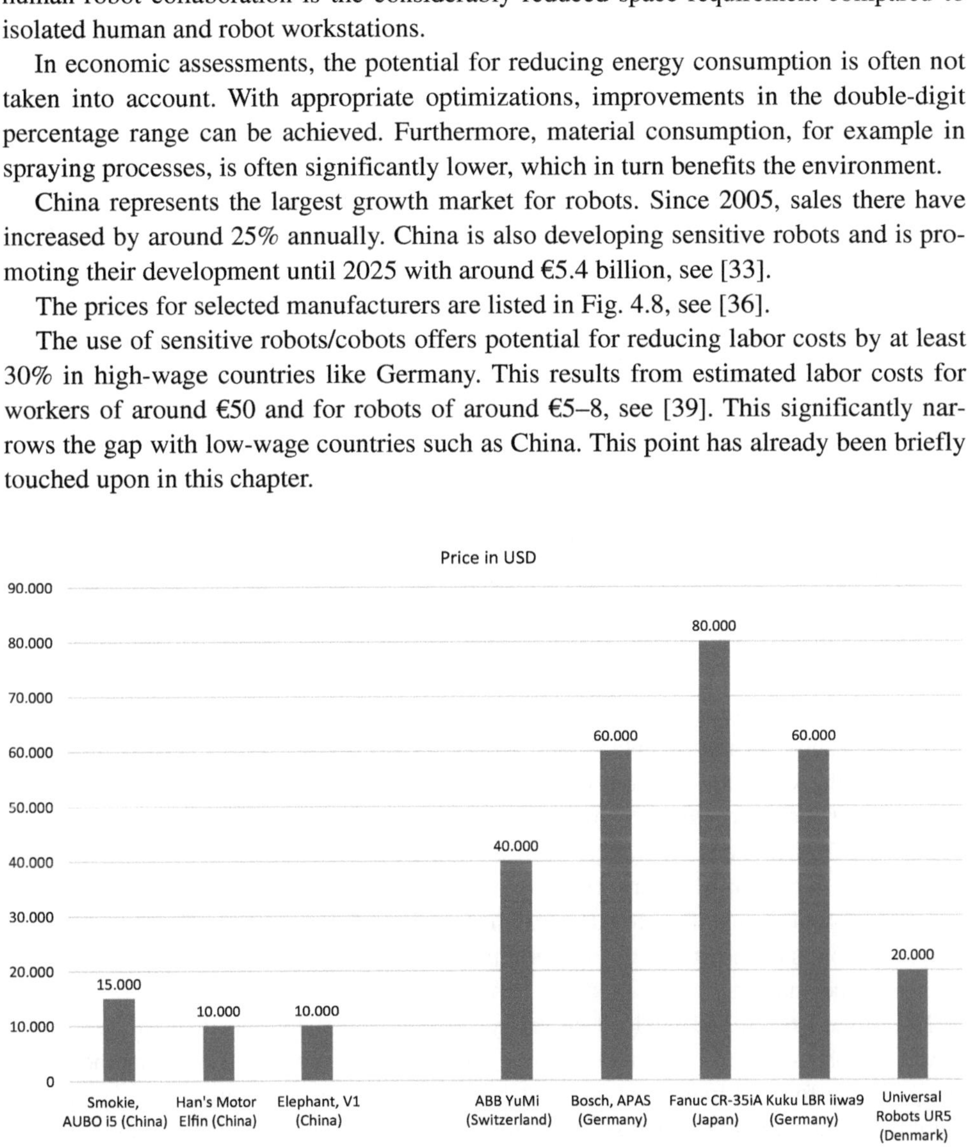

Fig. 4.8 Prices for sensitive robots

Currently, sensitive robots/cobots operate in cooperation with humans. This can increase output by 50%, [59] although this is, of course, highly dependent on the specific application.

Another aspect is that the more ergonomically favorable workplaces can significantly reduce employee sick days. Companies report a 30% improvement, along with increased employee satisfaction and longer employment of production workers due to age-appropriate workplaces, see [39]. Employee workstations are made significantly more attractive by such technologies.

4.6.4 Risks

The early involvement of the company's own works council is certainly a success factor. As already mentioned, it is strongly advised against reducing staff. This applies to all topics and technologies discussed here.

In addition, there are also technical risks. Meeting safety requirements is often associated with considerable effort and, in some cases, must be carried out individually for each robot deployment. Professional associations typically provide support in this area. Due to the existing risk of injury to employees, components with sharp edges or hot surfaces are currently excluded from use with sensitive robots, see [31].

The division of labor between humans and robots must be redesigned when they are deployed. It is important to take both sides (employees and the company) into account. The same applies to human-robot interaction (e.g., control by touch, gestures). Employee training measures should also not be forgotten. They are also an important success factor when introducing such systems, see [31].

Despite all the enthusiasm for sensitive robots/cobots, especially in connection with AI approaches, it is important to keep an eye on implementation efforts. A rollout requires the deployment of significant resources. Error rates of such solutions in the low percentage range are not acceptable and prevent practical use. Here, too, it is important to build up the know-how of your own employees at an early stage. The involvement of external staff and thus external know-how represents a practical approach.

For so-called low-cost countries, such as India, automation solutions are often only suitable to a very limited extent. Here, it is necessary to establish appropriate solutions through alternative manufacturers (for example, from Africa). In general, it is not advisable to roll out such solutions globally in operational practice. Local conditions must be taken into account.

The limited reach and payload currently set boundaries for sensitive robots/cobots. However, robots can also be made mobile. Again, the combination with another technology in the form of intelligent automated guided vehicles could provide an answer. It should also be emphasized that sensitive robots/cobots do not represent an alternative to classic industrial robots, but rather open up further automation potential.

4.7 Additive Manufacturing

Additive manufacturing (colloquially also referred to as 3D printing) will change the production landscape. Since the processes are extremely flexible, they can be used to manufacture customized products. The restrictions known from subtractive (i.e., drawing and machining) manufacturing do not apply to this approach. The processes have long outgrown prototyping and toy manufacturing. Additive manufacturing processes now produce, across industries, durable and market-ready products from small to large series. The process itself is not truly new at its core. Automotive manufacturers such as BMW have been working with the technology since the 1990s.

As early as 2014, according to a PricewaterhouseCoopers survey, around 11% of the companies surveyed had switched to series production using 3D printing, see [7]. By 2019, it was already one in three companies [60]. Many well-known manufacturers are among the users of the technology, such as General Electric (jet engine parts), Siemens (gas turbine parts), BMW, Audi (toolmaking), Luxexcel (lenses for LEDs), Airbus Group (parts for Airbus aircraft and Ariane 6 rocket). Selected application examples can be found in Fig. 4.9. The figure includes, among other things, a hip implant.

Thus, examples can be found from virtually all industries, including healthcare. However, these manufacturing approaches will not only fundamentally change product design and production, but also the entire logistics sector.

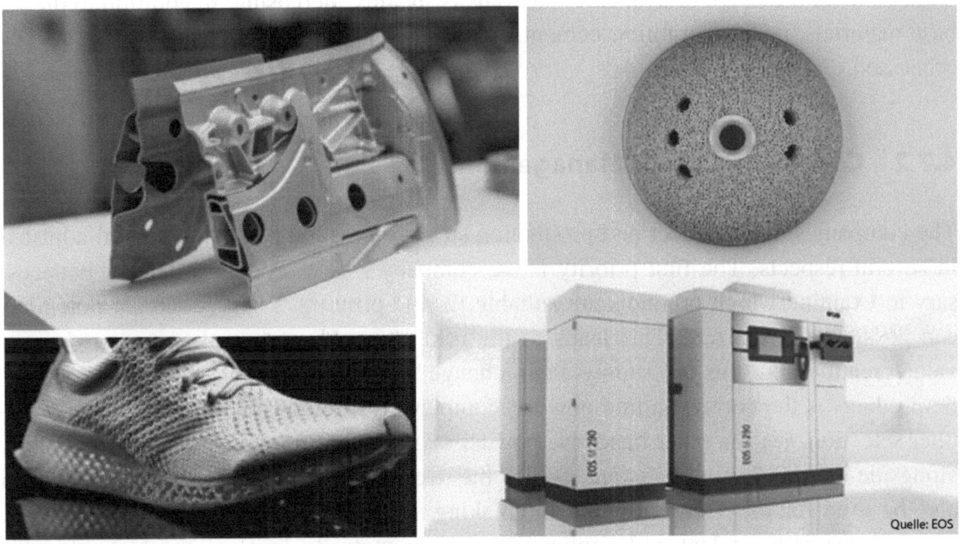

Fig. 4.9 Examples of additive manufacturing

4.7.1 Technical Description

Additive (generative) manufacturing processes refer to the creation of a component by adding material layer by layer, in contrast to classic subtractive processes (drilling, milling, turning, etc.). The material is applied by laser-based melting of powder. Control is handled by software. This manufacturing approach thus minimizes material usage. The advantages of using 3D printers are in the area of complex geometries or conventionally difficult-to-manufacture parts. This also makes it possible to produce parts that were previously either technologically or economically unfeasible to manufacture, or only with considerable effort. Various laser-based approaches exist for product manufacturing. The most common processes are:

- Melting processes,
- Sintering processes,
- Stereolithography.

Melting processes have the major advantage of low cost and smooth surfaces. On the other hand, there is a limited number of materials that can be processed. The advantages of the sintering process lie in the high mechanical strength of the objects produced and the large number of materials that can be processed. The advantages of the stereolithographic process are in the high surface quality.

In general, the range of available materials is steadily increasing. In addition to plastic base materials, resins, ceramics, cement, glass, and a wide variety of metals can also be processed.

4.7.2 Consequences for Management

The company's own product and production strategy must be reconsidered and adjusted in several respects. The first priority is the company's own products. Here, it is necessary to examine which products are suitable for 3D printing. What economic potentials can possibly be realized? This leads to the next point: How does the company's own value creation and thus the supply chain change? Since the starting material for metallic products is the corresponding powder, completely different possibilities arise in supplier selection. On the other hand, the powder is an important quality characteristic and, alongside the monitoring software, is also the decisive factor. However, the process can also be used for the very important toolmaking sector, as the example of Audi shows. The industry-independent and application-specific maturity level of the technology is shown in Fig. 4.10.

The shift towards additive manufacturing can indeed happen very quickly, as the example of the US hearing aid industry shows. This example was already introduced in Chap. 2. Companies that did not recognize the signs of the times did not survive, see [7].

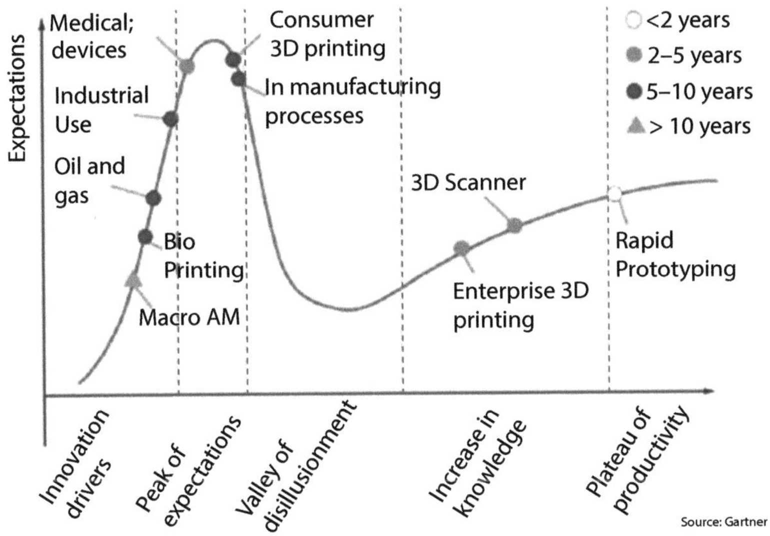

Fig. 4.10 Gartner Cycle for Additive Manufacturing

The advantage here was less about saving individual process steps, but rather the entire process chain could be optimized, resulting in significantly better and more consistent quality. Further examples could affect the healthcare sector in the coming years. Dental implants are currently manufactured in specialized laboratories based on impressions, which is a complex process. Using 3D cameras and 3D printers, dentists could also simplify the process chain, make laboratories redundant, and provide the long-awaited dental prosthesis to the suffering customer more quickly.

Regardless of the great potential, the high investment costs must not be neglected. These can easily reach into the six- to seven-figure euro range. The size of the component to be manufactured is a central cost factor. These investments can be mitigated in the initial phase through contract manufacturing (i.e., contract printing). If the forecasted potentials can also be realized in practice, the next step can be taken in the form of acquiring one's own 3D printers. Furthermore, there are industry-specific peculiarities to consider. For example, in the aerospace industry, extensive product certifications are required, which can only be achieved through destructive testing. These are necessary because, unfortunately, not all materials in powder form have the same material properties as those produced by conventional manufacturing. In general, the hype surrounding 3D printing has given way to a more realistic assessment. In addition, continuous innovations are leading to corresponding changes in the manufacturer market. Nevertheless, the market is very promising, see [64].

Moreover, there is currently an upper limit of about 1 m in component dimensions. Larger parts therefore cannot be manufactured.

4.7.3 Economic Relevance

The market surrounding additive manufacturing is experiencing very strong growth. From 2017 to 2022, it more than doubled globally from 8 to 18 billion USD. This market is divided into services and products. In 2022, an increase of over 18% was achieved. For the USA alone, an annual growth rate of 22% was reached by 2023, see [1].

Siemens was able to reduce repair times for gas turbines by around 90%. This was made possible by a changed design and subsequent additive manufacturing, see [17]. Siemens' competitor General Electric achieved a similar result. The company produces around 45,000 parts for jet engines per year. By reducing the number of components from 20 to 1, manufacturing costs were reduced by 75%, see [7].

Airbus has also additively manufactured brackets for the construction of the Eurostar E300 telecommunications satellite. As a result, the components could be made about 40% stiffer and 35% lighter. At SpaceX, the focus was on a significant reduction in manufacturing times for their rocket engine chambers. These could be reduced from several months to 2 days, see [25]. The ArianeGroup is also simplifying the production of its Ariane 6 engines using 3D printing. The number of components for the injector head is reduced from 248 parts to 1, which means halving the manufacturing costs. Similar to SpaceX, ArianeGroup was able to reduce manufacturing times from about 3 months to 35 h, see [10].

Spare parts production is also being significantly changed by additive processes. Until now, spare parts had to be available for a long time, which incurred corresponding costs. The required parts can now be "printed" on site as needed. 3D printing can significantly reduce storage costs in particular. In the future, it will be sufficient to archive 3D design drawings. Thus, this is also about the fundamental change in business relationships and production processes. Ultimately, it is simply about control of the value chain.

4.7.4 Risks

There is a risk that platforms will emerge to handle the exchange between developers, manufacturers, and customers. They could then also set standards. Manufacturers would then suddenly become mere suppliers in a newly designed supply chain.

Additive manufacturing approaches naturally make it much faster, better, and easier to copy products. The already significant problem of product piracy is further exacerbated by 3D printers. While this problem is already being recognized to some extent, applicable solutions are not yet widely available.

The rapid shift to additive manufacturing shows that waiting too long and clinging to old ways can have extremely fatal consequences, up to and including existential threats. The example of the US hearing aid industry may be extreme, but it illustrates the trends and potential dangers.

On the other hand, it should not be forgotten that when using several 3D printers to produce the same product, the quality of the results is not necessarily identical. It is therefore important to invest in software for online quality monitoring and to select the manufacturer accordingly, among other things.

Additive manufacturing will also change the logistics industry in the medium to long term. With 3D printers, spare parts can be "printed" on site in authorized workshops and do not need to be transported by logistics companies. This could allow these workshops to further differentiate themselves from independent workshops and bring the lucrative business "into the workshops"—to the detriment of logistics providers.

Despite the current wide range of applications, this technology still has its limitations. Large-scale use in series production is currently only possible in isolated cases due to the high costs and long "printing times" and is only economically attractive in certain situations. However, especially in the area of small series, this approach represents a very interesting alternative to conventional manufacturing processes. On the other hand, additive manufacturing enables unmanned production or at least individual shifts. This has a significant impact on the cost structure and at the same time puts long lead times into perspective.

4.8 Platforms

Platforms (especially IoT platforms) and digital ecosystems are much-used terms. Unfortunately, there is once again no clear definition for either term, nor a clear understanding of what such a solution actually is. On top of that, the market is highly dynamic. Furthermore, individual providers address different customer groups. This ranges from B2B to B2C. As a result, the solutions offered are sometimes difficult or even impossible to compare.

Virtually all major companies such as Amazon, Bosch, Siemens, IBM, SAP, Microsoft, Google, Intel, and Cisco have developed appropriately named solutions over the past few years. The term IoT operating system is also used synonymously for platforms in the B2B environment. In addition, various startups are trying to establish themselves in the market, especially with specialized solutions.

Regardless of interpretation and possible maturity, they abstractly offer services. The most well-known example is certainly Amazon. Having started with selling books online, the company now offers a wide range of services, including music, groceries, and the ability to store data in their cloud and process orders for companies via the cloud. All such solutions are cloud-based. Local implementations are sometimes possible. Thus, platforms are closely linked to cloud approaches.

The most relevant providers of platforms, in alphabetical order, are: Alibaba, Amazon, Google, Microsoft, and Siemens. All German automotive manufacturers have entered into partnerships with at least one of the aforementioned providers [3], [5]. It is therefore not surprising that in 2021, approximately 82% of all German companies used cloud services [61].

In the cloud provider market, neither a German nor a European company has managed to prevail. To regain data autonomy, Gaia-X was launched at the EU level in 2019. On the national German level, Catena-X and Manufacturing-X were established as concrete implementations. These approaches are considered flagship projects of the federal government. However, the current reality is quite different. Most key players are pursuing their own solutions within closed ecosystems [61]. Nevertheless, these platforms are becoming increasingly relevant.

4.8.1 Technical Description

Analysts such as Forrester attest that IoT platform providers are technologically mature, see [4]. The following section differentiates between B2B and B2C platforms. The latter can, in the broadest sense, also be described as marketplaces. The wide range of offerings can best be illustrated by concrete examples. The topic of platforms is always linked to the topic of cloud, see Chap. 5.

The prime example is Amazon. The company currently offers an extremely wide range of goods for both end consumers and commercial customers via its own platform. At www.amazon.de/business, business customers can register free of charge. Since December 2016, companies have had access to over 100 million different items for purchase. Approval workflows, account management, and multi-user accounts are available. For sellers, Amazon offers a simple way to offer goods and expand their customer base with little effort and risk.

The three main reasons for using the platform are good user guidance, a wide product selection, and low costs, see [9]. Of course, Amazon often offers free shipping, as is already the case with many products in the B2C sector. For Amazon, the number of companies offering their goods and services via the platform is important. This enables network effects and the provision of further data-based services.

Other providers focus on selected customer segments. Well-known examples are hotel portals or travel booking portals. Technically, these are web-based solutions. They are therefore conveniently accessible via mobile devices. Such platforms or digital marketplaces consist of:

- a communication and interaction platform,
- a trust center, including payment functionality and identification between buyer and seller,
- physical logistics.

In many cases, platforms offer all functionalities. For example, eBay, through its subsidiary PayPal, has also taken on the role of a payment service provider.

In contrast to Amazon, companies such as IBM, Siemens, or SAP offer their customers very complex services in the B2B sector. This includes, for example, big data applications, integration and control of an entire factory or its machines, including order management and planning. Such platforms are characterized by their functionality, i.e., the services offered. These are sometimes provided by the respective operator or by partners. Even with these complex services, the advantage for partners is the ability to reach new customers or customer segments easily and with little effort via the platform. Figure 4.11 provides an insight into how the individual platforms are presented. In recent years, it has become apparent that cloud-based solutions from companies such as Siemens or GE have not managed to achieve a relevant market position comparable to American providers, with corresponding consequences for GE. Siemens has made a fresh start here and is continuing its path into the digital world. Mendix is currently the most promising solution.

The technical basis for such solutions is cloud-based approaches. This means customers do not have to build and operate expensive and time-consuming IT infrastructure. For example, BMW and Schaeffler have entered into strategic partnerships with IBM and its Watson solution.

From what has been said so far, some commonalities emerge among the very heterogeneous platform offerings—the broad range of services, partners, and thus the size of the ecosystem. Both are decisive criteria for the success or failure of platforms. This applies regardless of whether it is a B2B or a B2C platform.

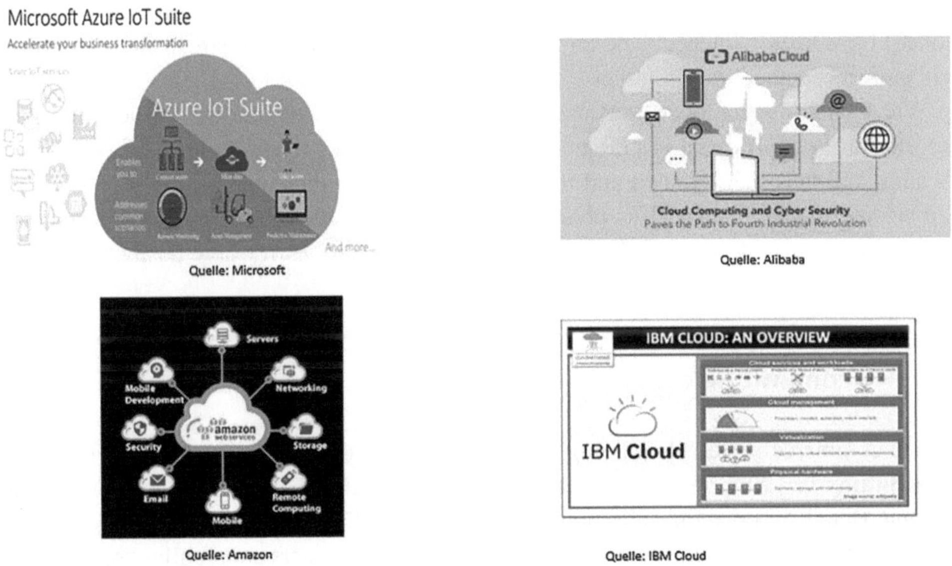

Fig. 4.11 IoT platforms for industrial customers

4.8.2 Consequences for Management

With platforms, it is important to distinguish between the two sides: the customer, i.e., the user of such services, and the supplier/partner. First, the supplier/partner side.

As already mentioned, B2B platforms make it possible to quickly access new customer segments with little financial effort. The financial risk is also manageable. On the other hand, there is a loss of direct customer access. The platform, so to speak, inserts itself between the manufacturing company and the end customer. As a result, important information about customer behavior and their wishes is lost. Pricing can be determined less and less by the manufacturer. The more powerful a platform is, the more it is able to determine the price and other conditions. This can even lead to the company becoming completely invisible in the market. In this way, the platform then dominates the respective market.

Participation in platforms for industrial companies should therefore be carefully considered. Building your own platform involves considerable financial expenditure and thus also risks, as the case of GE shows. The situation is different for consulting or software companies. They can use this as a simple way to access new markets. Regardless of economic relevance, appropriate services can increase reputation and demonstrate expertise in relevant topics.

In general, platforms have become indispensable in economic life, including their significance. In order to counter the dominance of American and Chinese companies with a corresponding European and German solution, appropriate initiatives (as already mentioned) have been launched. The coming years will show to what extent these solutions will gain acceptance in the market.

When building platforms, the user context (i.e., the need) and the attractiveness of the offering are of crucial importance. In the marketplaces of the future, the customer's location and thus their context and the situation in which the customer finds themselves can be determined, and specific, personalized offers and services that match their current needs can be provided, see [27]. This opens up opportunities for new business models, see Chap. 8.

4.8.3 Economic Relevance

Online commerce is steadily gaining in importance. While sales in Germany were still at €1.25 billion in 1999, by 2013 they had already reached €33.1 billion. According to the German Retail Association, sales continued to rise to €97.4 billion in 2022. According to the industry association BITKOM, nine out of ten internet users shop online, see [27]. In the B2C business, hardly anything works anymore without digital marketplaces, not to mention growth rates. The same applies to the search for information before a purchase. Here, too, more and more customers are turning to digital platforms. This trend was significantly reinforced by the coronavirus pandemic.

According to Frost&Sullivan, online B2B sales grew to around US$6,700 billion worldwide in 2020, while the B2C market reached US$3,200 billion. Thus, B2B sales would be twice as large as B2C sales, see [9]. Amazon has so far primarily operated in the B2C market. However, with Amazon Business and Amazon Web Services, the company is making a strong push into the attractive B2B market. The same applies to eBay Business Supply. As a result, strong companies are entering a new market environment and threatening existing providers. The China-based company Alibaba is also intensifying its global activities and is among the TOP3 in the industry. More on this in Chap. 10.

In general, as the economic relevance of platforms increases, so does data traffic. In 2013, this was still at 1.8 zettabytes, rising to 2.4 in 2014, 3.2 in 2015, and around 5.3 zettabytes in 2017, see [8]. The five largest platform operators worldwide (Google, Apple, Facebook, Amazon, and Microsoft) had a market capitalization of €4.5 trillion in 2019, compared to Germany's gross domestic product of €3.44 trillion, see [63].

Another manifestation of platforms is the use of this technology for innovations in the form of so-called "open innovations." American automobile manufacturers are already taking this approach to realize product developments at the customer, with the customer, and for the customer.

4.8.4 Risks

Possible risks have already been hinted at. Providers can easily become suppliers. Examples include the music industry with services such as iTunes, which have outpaced established providers like Sony Music and partially relegated them to the role of suppliers. In addition, the CD has become practically obsolete, as music is now available anytime and anywhere through these new services. The same fate befell the DVD. Buyers hardly notice that they no longer purchase music or movies from such services, but instead acquire only a right of use—or even see this as a positive. For example, in 2012, about 18.7 million downloads of individual tracks and albums were made in Germany, and in 2013 about 97.1 million via the internet and corresponding platforms. However, the number of downloads has been declining for years and amounted to just under 50 million in 2023, see [34]. In contrast, the market for music streaming services is growing strongly worldwide. In 2020, the increase for all providers was in the double-digit millions, see [62]. As a reminder, iTunes launched in Germany in 2004 and Spotify in 2012, see [21].

Analogous scenarios, albeit in a milder form, are provided by booking platforms such as Airbnb for accommodations. Airbnb does not own a single hotel, yet offers accommodation in over 191 countries and more than 34,000 cities. In Germany alone, there are more than 28,000 apartments. Today, Airbnb is estimated to be worth around US$10 billion, while Hyatt, as one of the largest hotel groups worldwide, is valued at US$8.4 billion. According to market experts, Airbnb alone took away around 1 million overnight stays from hotels in New York in 2013, see [21]. The company's advantage is that it does not have the high fixed costs of traditional hotel chains. Incidentally, the company was founded in 2007, which again demonstrates the strong market dynamics; more on this in Chap. 6.

References

1. Niro Bathija, Warum die einstige Hpye-Branche 3D-Druck wieder interessant wird, Focus online, 29.07.2023.
2. Michael Bauer, Industrieroboter – Neue Fertigungstechnologien aufgrund roboterbasierter Automatisierung, Vortrag, http://www.produktionstechnik.at/fileadmin/wpk/oc/2014_Vortraege/2c_Schweissen_I/6_Neue_Fertigungstechnologien_im_Bereich_roboterbasierender_Automatisierung.pdf, aufgerufen am 22.07.2017.
3. Martin Bayer, Sicherheitsfalle Digitalisierung: Cyber-Kriminelle verursachen Milliardenschäden, Computerwoche, 26.07.2017.
4. Kurt D. Bettenhausen, Stefan Kowalewski, Thesen und Handlungsfelder: Cyber-Physical Systems: Chancen und Nutzen aus Sicht der Automatisierung, VDI/VDE, 4/2013.
5. Manfred Bremmer, Forrest Wave IoT-Plattformen: Marktführer, „Strong Performer“ und Anwärter, Computerwoche, 30.11.2016.
6. Manfred Broy, Christian Prehofer, Hermann Engesser, Das Interview: Digitalisierung und die Rolle der Informatik in der Anwendung und Forschung, Informatik Spektrum Band 39, Heft 6, 12/2016.
7. Thomas H. Davenport, Julia Kirby, Dein Freund der Roboter, Harvard Business Manager, September 2015, Seite 22–31.
8. Richard D'Aveni, 3D-Druck vor dem Durchbruch, Harvard Business Manager, Juli 2015, Seite 20–29.
9. Markus Dettmer, Janko Tietz, Der Sieg der Algorithmen, in Digitale Revolution – Wie die Internet-Ökonomie unser Leben verändert: Ein SPIEGEL E-Book, Mai 2014.
10. Marcus Fost, Robin Eckardt, B2B-Plattforme: Wie sich Verfolger gegen Amazon Business wappnen müssen, e-tailment, 31. Januar 2017.
11. Detlef Flach, Raketendüse aus dem Drucker: Rakete Ariane 6: 3D-Druck reduziert Kosten um die Hälfte, Computerwoche, 18.07.2017.
12. Thomas Geiger, Ein Autoleben lang: Mit Updates und Upgrades will BMW den Kaufprozess revolutionieren, Automobilwoche, 24.07.2017, Seite 9.
13. Ute Holtmann, Roboter werden den Handel verändern, in Lebensmittel Zeitung, LZ 39, 29. September 2017, Seite 45.
14. IBM, Wer weniger Zeit mit Wartungsproblemen verliert, kann pünktlicher starten, https://www.ibm.com/cognitive/de-de/outthink/airlines-with-watson.html, aufgerufen am 29.07.2017.
15. IBM Können Satelliten dem Weingut E. & J. Gallo helfen, Wasser zu sparen und besseren Wein zu machen?, https://www.ibm.com/cognitive/de-de/outthink/ejgallo-with-watson.html, aufgerufen am 29.07.2017.
16. Masaaki Imai, KAIZEN – Der Schlüssel zum Erfolg im Wettbewerb, Econ-Verlag, 1986.
17. Henning Kagermann acatech (Hrsg.), Innovationspotenziale der Mensch-Maschinen-Interaktion, April 2016, http://www.acatech.de/fileadmin/user_upload/Baumstruktur_nach_Website/Acatech/root/de/Publikationen/Stellungnahmen/acatech_IMPULS_Mensch-Maschine-Interaktion_WEB.pdf, aufgerufen am 08.08.2017.
18. Ulli Klenk, Von dieser Technik träumte jeder, Harvard Business Manager, Juli 2015, Seite 34–35.

19. Werner Kurzlechner, Ratschläge von McKinsey: 8 Use Cases für Künstliche Intelligenz, CIO, 14.07.2017.
20. Wolfgang Kobek, Big Data Analytics für IoT-Szenarien: Nicht Wunsch, sondern Realität, Computerwoche, 27.07.2017.
21. Tal Leder, Kampfroboter und KI – In Israel wird Science-Fiction Realität, ntv, 29.07.2023.
22. Pavel Lokshin, Das große Krabbeln der Roboter, Zeit online, 13. Februar 2014, http://www.zeit.de/wissen/2014-02/termiten-bauroboter/komplettansicht?print, Aufgerufen am 27. Juli 2017.
23. Armin Mahler, Thomas Schulz, Digitale Revolution – Wie die Internet-Ökonomie unser Leben verändert: Ein Spiegel E-Book, Mai 2014.
24. Klaus Mainzer, Künstliche Intelligenz – Wann übernehmen die Maschinen?, Springer, März 2016.
25. Andrew McAfee, Erik Brynjolfsson, Besser entscheiden mit Big Data, Harvard Business Manager, 4/2014.
26. Thomas Matyssek, Geschäftsmodelle im Internet der Dinge, in Daniel Schallmo (Hrsg.) Digitale Transformation von Geschäftsmodellen, Springer Gabler, November 2017, Seite 159–177.
27. Christiane Kaiser-Neubauer, Robo-Advisor-Markt: Raus aus der Nische, Süddeutsche Zeitung, 12.8.2020.
28. Philip Paul, Die Zukunft der Raumfahrt im Blick – SpaceX, https://www.3d-druck.de, 20. Januar 2016.
29. Claudia-Linnhoff Popien, Tim Frohwein, Wie die menschliche Intelligenz erweitert werden kann, Digitale Welt 3/2017, Seite 17–21.
30. Claudia Linnhoff-Popien, Michael Zaddach, Andreas Grahl (Hrsg.) , Marktplätze im Umbruch: Digitale Strategien für Services im Mobilen Internet, Springer Vieweg, April 2015.
31. Michael E. Porter, James E. Heppelmann, Wie smarte Produkte den Wettbewerb verändern, Harvard Business Manager, 12/2014.
32. Projekt zur dezentralen Energieversorgung: Cloudlösung ermöglicht zentral Überwachung und Steuerung, elektro.net, 26. Juni 2017.
33. August-Wilhelm Scheer, Nutzentreiber der Digitalisierung – Ein systematischer Ansatz zur Entwicklung disruptiver digitaler Geschäftsmodelle, Informatik Spektrum, Band 39, Heft 4, August 2016, Seite 275–289.
34. Marc-Andre Weber, Zahlen, Daten, Fakten: Mensch-Roboter-Kollaboration, Institut für angewandte Arbeitswissenschaft e.V. (ifaa), März 2017.
35. Sabine Spinnarke, KI auf dem Shopfloor: So wird künstliche Intelligenz in der Produktion eingesetzt, Produktion, 3. Mai 2017.
36. Max Roser, The brief history of artificial intelligence: The wold has changed fast – what micht be next? OurWorldinbData.org, Dezember 2022, publiziert in HBM 07/2023.
37. Alec Ross, Die Wirtschaftswelt der Zukunft, Plassen Verlag, 2016.
38. Gerd Scholz, Daimler sucht die meisten KI-Experten, Automobilwoche, 24.07.2017, Seite 19.
39. Soren Skov, Maersk Q2 2017 report, 16 August 2017, http://files.shareholder.com/downloads/ABEA-3GG91Y/5070505423x0x954063/4803DE66-B269-4730-8A92-C0D59BD7EE28/Presentation_Q2_2017.pdf, aufgerufen am 28.08.2017.
40. Seitenblick: China – Droht die Roboterblase?, Wirtschaftswoche, 27./30.6.2017, Seite 14.
41. Vermögensverwaltung der Zukunft, Einblicke, 02.2017, Seite 19–26.
42. Was denkt sich die KI?, Technology Review, August 2017, Seite 31–34.
43. Horst Wildemann, Produktivität durch Industrie 4.0, 24. Münchner Management Kolloquium, 14./15. März 2017.
45. Datenmenge explodiert, https://www.iwd.de/artikel/datenmenge-explodiert-431851, 07. Juni 2019.

46. Spiegel, Zahl der Onlinebanking-Nutzer steigt um drei Millionen, 22.03.2021, dpa.
47. bitkom, Pressemitteilung 203 Mrd. € Schaden pro Jahr durch Angriffe auf Deutsche Unternehmen, Simran Mann, Andreas Streim, 31. August 2022.
48. Data Lake for Dummies und Architekturkonzepte zur Datenverwaltung in Data Lakes, Dissertation, Corinna Giebler, 8.12.2021, Univ. Stuttgart, Institut für Parallele und Verteilte Systeme.
49. Martin Heller, Was ist ein Data Lake, Computerwoche, 06.05.2022.
50. Alan R. Simon, Data Lake für Dummies, For Dummies, 1. Edition, 25. Juni 2021.
51. Time Cole, Erfolgsfaktor Künstliche Intelligenz, Hanser Verlag, 2020.
52. Jürgen Hill, Wer braucht den digitalen Zwilling wirklich, Computerwoche, 30.12.2012.
53. Ralf Otte, Künstliche Intelligenz für Dummies, Wiley–VCH Verlag 1. Auflage, 2019.
54. Ralf Otte, Maschinenbewusstsein, Campus Verlag, 2021.
55. Daniela Hoffmann, Transparente Kette, Automobilproduktion 04/2022.
56. BITKOM, Achim Berg, April 2019: Industrie 4.0 – jetzt mit KI.
57. Anne Steinbach, „Künstliche Intelligenz ist der Wachstumstreiber", Springer Professional, 11.06.2019.
58. Dr. Barbara Stumpp, Smarte Sensoren für die Digital Factory, Automation & Digitalisierung, 30.04.2020.
59. Susanne Nördinger, Nur ein Hype oder ein Kassenschlager? Mensch-Roboter-Kollaboration: Wo sich Cobots wirklich lohnen, Produktion, 11.02.2020.
60. Pressemitteilung, Deutsche Industrie setzt auf 3D-Druck, Nina Paulsen, Angelina Marko, 14. Juni 2019.
61. Automobil Produktion, „So steht es um die Automotive-Cloud Catena-X", 9. Juni 2022, Claas Alexander Stroh.
62. Stefan Mayrhofer, 13.07.2020, „Apple Music: 25 % Marktanteil in rasant wachsendem Sektor".
63. BMWK, www.bmwk.de, Digitale Plattformen, aufgerufen am 07.01.2023.
64. Christof Huth, Bernhard Langefeld, Investitionen in die additive Fertigung, Roland Berger, 19.05.2022.
65. VDMA, Nachfrage nach Medizinrobotern steigt kräftig an, medizin&technik, 28.10.2020.
66. Jürgen Hill, BMW testet nächste Generation humanoider Roboter, CIO, 09.08.2024.
67. BITKOM, Künstliche Intelligenz, 2020.
68. Studie: Generative KI kann zum Produktivitätsbooster werden 13. Juni 2023 | Press Release, Philipp Hühne.
69. Markus Becker, Patrick Beuth, Alexander Demling, Christoph Giesen, Anton Rainer, Marcel Rosenbach, Bernd Zand, Wettlauf der Gehirne, Spiegel, 10/4.3.2023.
70. Paul Wellener, Steve Shepley, Ben Dollar, Manufacturing goes digital: Smart factories have the potential to spark labor productivity, Deloitte, 16. September 2019.…
71. Steffen Herbold, Data-Science-Crashkurs: Eine interaktive und praktische Einführung, dpunkt Verlag, 2022.
72. Michael Kroker, Weltweite Datenmenge soll bis 2025 auf 175 Zetabytes machen - 8 mal so viel wie 2017, Wirtschaftswoche, 27.11.2018.

Technologies That Evolve 5

The previous chapter described technologies with disruptive characteristics or transformative potential. However, in addition to these revolutionary looking technologies, there are others as well. These will also change the economy and businesses—albeit in a somewhat more subtle manner. These technologies were already briefly introduced in Chap. 2 in the overall context. Now, the aim is to take a closer look at them.

As already mentioned in Chap. 2 when classifying individual technologies as disruptive, the grouping of technologies as evolutionary is, of course, also somewhat subjective. Strictly speaking, none of the technologies described here are "new"; they have been known for many years. Since these technologies are "only" evolutionary in nature, they are presented somewhat more briefly than their revolutionary counterparts. However, the same principle applies here: the combination of individual technologies is just as significant as the individual technology itself. In many cases, the benefits or impacts on companies are amplified as a result.

5.1 Cyber Security

Cyber security has become a central concern for companies, at the latest due to the political developments of recent years and as a result of the COVID-19 pandemic—or at least it should be. According to the Federal Office for Information Security (BSI), the situation continues to intensify and is now more critical than ever before. For example, at the beginning of 2023, hackers launched widespread attacks on German companies and government agencies. Increasingly, private individuals are also being affected by such attacks. The attacks themselves are diverse. The table below, Table 5.1, provides an overview of the "Top 10" types of attacks on industrial control systems according to the BSI, see [31].

W. Huber, *Industry 4.0 Compact—How Technologies Are Transforming Our Economy and Our Companies*, https://doi.org/10.1007/978-3-658-51029-9_5

Table 5.1 Top 10 Cybersecurity Threats

Number	Threat
1	Social engineering and phishing
2	Introduction of malware via external devices
3	Infection with malware via Internet or intranet
4	Intrusion via remote access
5	Human error or sabotage
6	Control software connected to the Internet
7	Technical malfunction and force majeure
8	Compromise of extranets and cloud components
9	(D)DoS attacks
10	Compromise of smartphones in the production environment

TOP 10 Threats

The topic of cyber security is highly complex. In the business context, a distinction can be made between

- cyber security of products (product security),
- of the company (IT security), and
- of production (operational technology security)

. Thus, it is not a single technology. Rather, it is a topic that encompasses society as a whole and the entire enterprise. As business models become increasingly digitalized and processes, products, and production facilities become ever more interconnected, the threat landscape continues to grow. The following provides a summary of the individual aspects of the term cyber security. This includes the connection and interaction between physical devices and digital systems. The spectrum thus ranges from PCs, servers, cloud, robots in production, network components, sensors, smartphones/tablets, all the way to employees/customers/partners.

5.1.1 Technical Description

In the analysis, it is irrelevant whether your own IT systems are operated within the company or by an external service provider. The same applies to production systems, such as robots. The responsibility for the security of your own company always lies within the company itself. In general, the goal is to establish what is known as an Information Security Management System (ISMS for short). This system defines roles, responsibilities, rules, procedures, measures, and tools (such as IT systems) that are used to manage information security. It thus represents a systematic approach to building such a system. The bad news is that this is not a static process. The threat landscape and specific approaches to cyberattacks are constantly evolving. Therefore, it is necessary to establish a continuous improvement process. The good news is that the BSI has defined a corresponding baseline protection based on established international standards (ISO 207001). This baseline protection helps companies in setting up an ISMS. Analogous standards also exist for the American market, such as the Cybersecurity Framework (NIST for short). For specific topics such as OT security (i.e., cybersecurity around production) and a company's products, there are additional national and international norms and standards.

Regardless of whether you want to handle the topic of cybersecurity internally or outsource it, you need a central point of contact within your own company. In general, time is of the essence here. An ISMS is the current standard and is increasingly being demanded by customers. Therefore, at least for the initial implementation of an ISMS, outsourcing is certainly the best choice. For small and medium-sized enterprises (SMEs), this could also represent a general approach.

The topic of cybersecurity has three dimensions:

- People
- Processes
- Technologies (usually IT systems, but also machines and equipment in production).

People, processes, and technologies must be secured. Each company must decide for itself what level of risk it is willing to accept and where its "red lines" are. This determination must be made consciously. In this context, it is also necessary to define specific so-called attack vectors and corresponding countermeasures.

However, these three dimensions also help in defending against attacks or in reducing risks. Employees, for example, need to be made aware, and IT solutions (such as firewalls or antivirus programs) help with defense. Processes must be designed to be resilient. However, it must be emphasized that there is no such thing as 100% security. Thus, the previously mentioned residual risk remains.

The question to be answered is what makes your own company particularly valuable and thus which physical and non-physical assets need to be protected. Answering this question forms the starting point for implementation. The so-called CIA model (Confidentiality, Integrity, and Availability) serves as a guideline. Integrity refers to the immutability of data and information. This includes deleting or modifying data. Based on these three dimensions, risks and impacts in the event of damage must be defined. A so-called penetration test (i.e., a targeted hacker attack) provides initial clarity about the current situation. It also helps to verify the effectiveness of current countermeasures.

5.1.2 Consequences for Management

The question is not whether your company will be successfully attacked, rather when and what impact this event will have on your company. Thus, this is a real, not a hypothetical, threat. The consequences are not limited to the economic sphere, but often also result in reputational damage and thus a loss of trust. The theft of confidential information (i.e., customer and company data) is just one possible scenario. Production shutdowns lasting several weeks represent another scenario, with corresponding economic consequences. Therefore, this topic should be a "top management issue." Building a comprehensive cybersecurity system is therefore absolutely essential. As already mentioned, the starting point is to identify the risks to your own company, initiate mitigation measures, and continuously monitor their implementation. The primary focus here is on incoming and outgoing information flows and on your own employees (keyword: social engineering as a possible attack vector). Your own employees are responsible for about 50% of damages or incidents, see [11]. It is necessary to establish a business continuity management system as part of the ISMS and to anchor it within the company, in order to minimize cyberattacks and their impacts. The introduction of a comprehensive cybersecurity system is complex and also financially demanding. It is therefore not sufficient to simply install the latest antivirus software or run an awareness campaign. It is absolutely essential to build a resilient organization within your own company. Raising awareness plays a central role in this. All personnel in the company must be informed and sensitized to the topic, specifically regarding protection objectives and current threats. All technical measures are useless if this goal is not achieved.

5.1.3 Economic Relevance

The economic damage can be differentiated between indirect (loss of reputation) and direct (such as production downtime). According to bitkom, the economic damage in Germany amounted to €203 billion in 2021, with 9 out of 10 companies falling victim to data theft, espionage, or sabotage, see [25]. If, for example, a production facility is affected, it can indeed lead to a shutdown lasting several weeks. Securing your own company is therefore vital for survival.

The aforementioned ISMS helps to prevent attacks and thus damage to your own company, and it is a prerequisite for taking out cyber insurance. Cybersecurity costs money. That should be clear, but doing nothing will cost significantly more. Therefore, companies are well advised to invest in meaningful protection. As so often, your own employees play a central role in this.

5.2 Vertical and Horizontal Integration

The topic of vertical and horizontal integration, see Fig. 5.1, has been a concern for companies for many years. Evolved system landscapes often prevent appropriate integration. Interfaces must be created and maintained with great effort and ultimately become obstacles to innovation. However, without this integration, there is no continuity and no smooth flow of business processes, especially no digital continuity. As a consequence, data and information are lost, are inconsistent, or must be maintained manually or multiple times. The importance of integration from product development through design to production and sales has already been recognized in KAIZEN—that is, Lean Management, see [7]. It is therefore not enough to focus solely on technical aspects; the associated processes must also be aligned accordingly. Specifically, this means that production must be integrated into product development at a very early stage.

Another goal of both integration approaches is the timely (real-time) provision of information for people, systems, and machines. A special case here is certainly shop floor integration, i.e., between ERP and MOM (Manufacturing Operation Management)/MES (Manufacturing Execution System) or PLM and MOM/MES. Here, seamless part and product identification is also at stake.

5.2.1 Technical Description

As already noted, this is not exclusively a technical issue, but also a process-related—and thus human—topic, since business processes need to be connected. As so often, it is also always about competencies and thus responsibilities within the company.

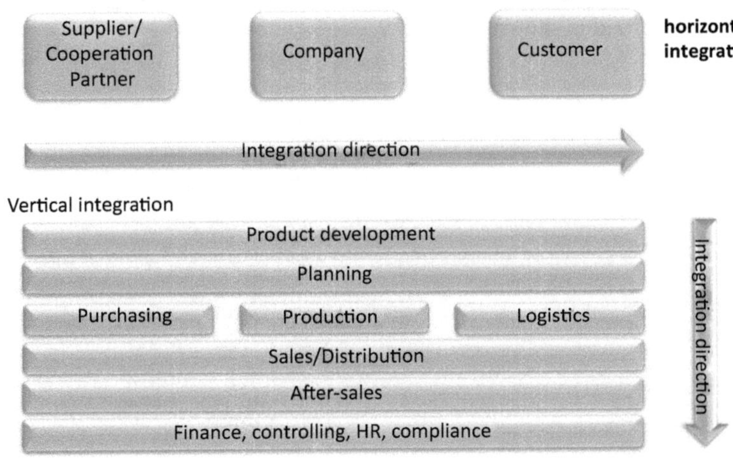

Fig. 5.1 Vertical and horizontal integration

The approaches to solving the problem are diverse and just as varied. However, due to the increasing digitalization of all areas, the issue is becoming ever more urgent and, unfortunately, ever more complex. Integration must be carried out down to the automation, i.e., machine level. The starting point for integration should be product development. The systems most affected by integration are:

- PLM (Product Lifecycle Management, i.e., product development),
- ERP (Enterprise Resource Planning, i.e., enterprise management),
- MOM (Manufacturing Operation Management, i.e., production and its control) including the machine level,
- MRO (Maintenance and Repair Order, i.e., maintenance).

In addition, the entire logistics chain must also be integrated. In horizontal integration, suppliers and customers are also affected and must therefore be integrated as well. It is of little use to have fully digitalized one's own company but not to take the supplier and customer side into account. As soon as a part or product leaves the company's own factory gates, it would thus sink into the analog "nirvana."

Technologically, the topic can be addressed via:

- Interfaces,
- Standard software, such as SAP or the common PLM providers,
- MSB (Manufacturing Service Bus),
- Cloud-based approaches.

to solve. There are also hybrid forms here. Standard software, such as SAP, is increasingly being offered via the cloud, and the same applies to the MSB approaches of various manufacturers, such as DXC, IBM, or the Fraunhofer Institute IPA in Stuttgart. An example of such an MSB in the cloud is shown in Fig. 5.2.

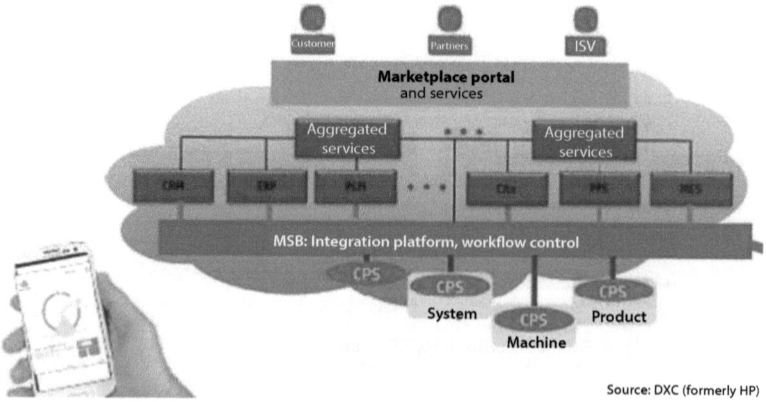

Fig. 5.2 Example of an MSB from the cloud

One of the central challenges here is master data management, i.e., well-maintained data, from the order, its feedback after successful completion, supplier deadlines, etc. In general, ERP-MOM integration is the simplest in terms of business and technology, while PLM-MOM is the most complex. Digital twin, digital product memory are just two of the buzzwords here. The connection to the automation level can be solved technologically, temporally, and economically most easily via the new OPC-UA (Unified Architecture) standard. Here, the question of which data are to be transferred from the machines to the MES and big data system and how they are to be processed must be clarified. Another example is the demand-driven supply of production with material (keyword so-called milk run, see [15]).

5.2.2 Consequences for Management

A precise quantification of the economic effects of the two types of integration is difficult to present, as the initial situations of companies often differ greatly. However, all German automobile manufacturers are making great efforts in this direction. Rapid response to market changes is no longer possible without an integrated approach. Without up-to-date information on production status, a production network cannot be managed efficiently. Thus, the topic must also, or above all, be considered from a strategic perspective.

The technical side, however, is only one side of the "coin." The greatest challenge is the standardization and integration of processes and thus of the individual departments. People from different departments must work together effectively. This starts with product development. Design to Manufacturing, i.e., product design taking into account economical production, should actually be state of the art. However, there is often still considerable potential for improvement here, as the collaboration between the individual areas is not fully practiced or only superficially, without generating real economic added value.

However, the ever-increasing technical integration also increases the risk of cyberattacks. Due to real-time requirements, production needs different protection mechanisms than, for example, ERP or office applications, as already described. Unauthorized access to data, machines, and systems must be prevented. Robots and SCADA systems in particular are attractive targets here, as their protection is often neglected. This was already pointed out in the previous section.

5.2.3 Economic Relevance

The two types of integration provide a holistic view, including of the entire supply chain, usually in real time as well. This forms the basis for decisions—based on the current situation and not on planned or outdated actual data. This allows, for example, inventory levels to be optimized, impending production bottlenecks to be identified in time, and maintenance measures to be optimized, thereby improving machine uptime.

Specifically, at the American motorcycle manufacturer Harley-Davidson, various measures (including vertical integration and standardization) reduced the lead time for producing a motorcycle from an average of 26 days to 6 hours, which corresponds to a reduction of about 90%, see [20]. Among German automobile manufacturers, the digitalization of the value chain—and thus vertical and horizontal integration—will increase from the current 20% to 80% in the coming years, see [16].

5.3 Blockchain

One way to implement horizontal integration is through so-called blockchains. However, they have significantly more potential. Bitcoins are very popular users of a blockchain. More on that in a moment.

The technology first appeared in 2008. Since then, there have been a wide variety of use cases for blockchains. Digital currencies, such as Bitcoins, are just a few examples. The financial sector has been significantly transformed as a result, but the food industry has also benefited from the technology. In logistics, it has only been partially adopted so far.

5.3.1 Technical Description

Business transactions are based on information and trust. The faster and easier this information can be exchanged, the better for those involved and the higher the acceptance of the technology used. This is precisely where blockchain demonstrates its strengths.

A blockchain is an immutable register (also called a transaction book or ledger) used by all business participants. All transactions carried out are stored here, regardless of whether they involve tangible goods (cars, cash) or intangible assets (patents, copyrights, trademarks, digital artworks). As a result, all relevant information can be bundled in so-called smart contracts. Everyone involved in the network has access to the distributed ledger. The individual entries are immutable. If a transaction goes wrong or an error occurs, a new one must be created. The data generated in this process is stored in the form of a block. Immutability is ensured by so-called cryptographic hash functions. Transactions are encrypted in such a way that it is not possible to draw conclusions about the input data. On the other hand, individual transactions are uniquely identifiable. This allows their authenticity to be clearly determined.

All transactions are linked together. In this way, a chain is formed. In total, this creates a blockchain, that is, a chain of immutable transactions.

5.3.2 Consequences for Management

Blockchain technology certainly received a great deal of attention in 2017 and 2018, when very large sums were invested in the technology and its applications. The best-known representatives or applications of the technology are certainly cryptocurrencies such as Bitcoin or Ethereum (as already mentioned). These are virtual, unregulated, decentralized payment systems.

In addition, contracts can also be executed via blockchains (so-called smart contracts). Due to the specific security properties of blockchain technology, it provides an excellent basis for smart contracts. The special feature is that no trust is required between the contracting parties. Therefore, this approach is particularly interesting for the logistics and supply chain sector. However, it should be mentioned that smart contracts are not immune to manipulation. There have already been cases where smart contracts were manipulated. Despite the many application examples, the user potential of this technology is still in its infancy. Therefore, further use cases will certainly emerge in the future. Companies are therefore well advised to engage with the technology, especially in the use cases already known today, see the next paragraph.

5.3.3 Economic Relevance

Payments can be processed entirely digitally using Bitcoin. There is no government regulation involved. Bitcoins are also traded very speculatively on the stock market. As a result, the value of the digital currency fluctuates up and down. You therefore need strong nerves or a long-term perspective.

The use cases are diverse and range from

- supply chains
- healthcare
- public administration
- oil and gas
- insurance
- financial services
- travel and transportation industry
- manufacturing.

The list therefore covers more or less the entire industry. Additional use cases are continuously being added. This results from the advantages of this technology. A blockchain is immutable, it offers complete transparency for all participants while fully respecting the privacy of the participants.

A negative aspect is the extremely high electricity consumption. The high energy usage is leading to increasing criticism of the technology. In addition, the applicability and thus user-friendliness are certainly not optimal. Combined with the high energy consumption, the overall costs of a blockchain are certainly on the negative side. The lack of scalability must also be considered.

Whether the disadvantages or advantages outweigh for a particular company must be assessed on a case-by-case basis. Due to the wide range of use cases, the technology certainly belongs in the toolkit of every corporate IT department, but it should also be constructively discussed in the business units.

5.4 Autonomous Transport Units

Autonomous transport units, like sensitive robots, are examples of Cyber Physical Systems (CPS), that is, the skillful combination of hardware, software, and sensor technology. In the context of Lean, logistics costs are considered unavoidable waste that should nevertheless be minimized. One approach to achieving this is the use of autonomously operating and self-controlling transport units, especially in intralogistics (i.e., internal company logistics and material flow), see Fig. 5.3.

However, the application scenarios are not limited exclusively to industrial facilities; they are also found, for example, in hospital automation. Autonomous cars or trucks are not included here, as their complexity and thus the legal hurdles are significantly higher.

5.4.1 Technical Description

Autonomously operating transport units are based on driverless systems, which have existed for around 60 years. During the CIM era, such systems were widely introduced

Fig. 5.3 Driverless transport unit

in German factories. The major difference today is self-control (i.e., autonomy) and sensor technology. Such systems feature:

- Central control (master control),
- Communication component,
- Navigation system,
- Warning and safety device,
- other peripheral devices.

By appropriately linking the individual autonomous transport units, swarm-based or agent-based systems can be created. The largest such systems have over 500 vehicles in a warehouse, see [27]. Regardless of technological development, autonomous transport units force structuring and organization (e.g., clear definition of travel routes and parking spaces in production). In practice, less intelligent systems are also used, such as track-based systems that move along defined paths. In some cases, charging can also take place simultaneously via these paths.

5.4.2 Consequences for Management

The range of applications for autonomous transport units is very broad. It extends from moving small containers to picking transport orders and transporting loads weighing several tons. Applications can be found in the automotive industry, aerospace industry, steel industry, or paper production. As stacker systems, they can operate not only on the floor but also move loads in high-bay warehouses. As a result, forklifts and overhead cranes operated by humans are increasingly being replaced by their digital and autonomously operating counterparts.

However, the effort required to install such systems should not be underestimated. It is important to consider whether changes to the floors or the environment are necessary. Integration into industrial automation should be pursued. Further automation of production can be achieved through machine-to-machine communication (M2M). Here, intelligent parts (for example, equipped with RFID tags) communicate independently with the autonomous transport units about the required transport. This leads to a significant reduction in complexity in production control and greatly relieves the respective employees.

The sensors present prevent harm to people. If a person enters the safety area, an emergency stop is triggered. Depending on the flexibility of the system, operation resumes only after confirmation by an operator. Therefore, a largely static environment is most favorable for their use. Such conditions should be taken into account already during the planning phase. The higher the degree of autonomy of such systems, unfortunately, the higher the costs and the complexity of their implementation. As a result, track-guided systems are increasingly used in many industrial companies, as this minimizes costs and maximizes benefits. Before implementation, it is therefore necessary to weigh which systems are the most cost-effective. The spectrum ranges from fixed, rigid lines to self-controlling, autonomously operating transport units.

5.4.3 Economic Relevance

The pure product costs are well below €20,000 per unit, of course depending on the specific product, and can therefore also rise into the six-figure euro range. The business models of individual providers vary considerably. Some providers also offer a pay-per-use approach in the form of transported load in kilograms per kilometer traveled.

By digitizing logistics using autonomously operating transport units and digitizing the milk run (this refers to supplying individual machines and workstations within production with the required components and materials, see [15]), potential savings of around 30% can be realized, see [1]. Return on investment can be quickly calculated based on the personnel costs saved. In multi-shift operations, the economic effects add up, or material supply can also take place at night, which may not be feasible with personnel-based material supply for cost reasons. However, the increased maintenance effort, including the required spare parts and qualifications of maintenance staff, must not be overlooked.

5.5 Real Time Enterprise and Mobile Solutions

For around 10 years now, a significant transformation has been evident among German automotive manufacturers, for example. Real-time applications are no longer seen merely as a component of core processes. There is even talk of real time itself becoming a core process. It is therefore clear that this topic represents a blend of technical and business challenges. The individual business processes are profoundly affected. An important prerequisite for this is efficient, i.e., lean and thus integrated processes—see previous descriptions. However, real-time capability here is not limited to internal processes, but should also include suppliers/partners and customers. Such collaboration requires a standardized value creation network.

Increasingly, the term real-time analytics is used instead of real-time enterprise. In the banking sector, this has been a key business model for years in the buying and selling of stocks.

5.5.1 Technical Description

A real time enterprise requires timely information delivery on mobile devices. Frequently, customized solutions are used. However, more and more standard software providers, such as SAP (especially with S4/Hana) or Microsoft, are offering corresponding solutions. Standardization should be implemented for the end devices. In management environments, Apple devices are often used for security reasons (and certainly also for image reasons), whereas in operational production environments, Windows 10 devices or Android-based devices are used. The Bavarian and Austrian police also use Apple smartphones, see [29]. In industrial settings, the use of mobile devices is not limited to pure display functionality. Active control is increasingly taking place, for example in maintenance. Tablets are also being used more frequently here. Again, there is a division between the management and operational levels.

The topic is supported by new technologies such as in-memory computing, big data, and sensor technology. RFID is also increasingly developing into an important technology in this context. It enables the identification, localization, and thus control of parts, containers, and end products.

5.5.2 Consequences for Management

The transition to a real-time capable enterprise is a complex and lengthy undertaking. The transformation should not take place as a "big bang," but rather as a continuous process starting with the most relevant business processes. In total, dynamic, real-time-oriented value chains are created, which can be optimized according to various key performance indicators.

It is important in this transformation to identify meaningful information in order to provide the right information to the right people at the right time. In the context of Industry 4.0, this is an absolute necessity. As a result, there is very high flexibility in process management, enabling such companies to respond very quickly to market changes. This is also associated with cost savings. The drivers in this context should always be on the side of the business processes. The goal is not to do things faster, but to do them smarter and thus better. A prerequisite for this is, among other things, the identification and definition of relevant KPIs to measure process efficiency and thus recognize the drivers for business decisions.

5.5.3 Economic Relevance

The advantage associated with real-time capability in the form of increased speed is also reflected in correspondingly greater success and economic returns. In addition to the strategic aspect, there is also a direct correlation between cost and speed. The conclusion is that real-time capability thus represents a virtual corporate value.

5.6 Digital Factory

The digital factory has now become part of day-to-day operations. In the German automotive industry in particular, planning is carried out first in the virtual, i.e., digital, space before a single "stone" is moved in reality. Manufacturing processes are also first simulated digitally before being implemented in reality at great expense and effort. The advantages are significantly higher planning reliability, thus avoiding planning errors, lower modification costs, and at the same time shortened product ramp-up times. Overall, investment and, ultimately, manufacturing costs of products are reduced while risks are minimized.

5.6.1 Technical Description

The digital factory consists of a comprehensive network of digital models, methods, and tools. The starting point here is the product to be manufactured. Systemically speaking, we are in the product development phase and thus in the PLM system (Product Lifecycle Management). An important prerequisite for the digital factory is consistent data management. The more complex the product, the more difficult it is in practice to meet this requirement. From a data perspective, the starting point is the digital drawing, or better yet, the digital 3D object model of the product. CAD drawings have the major disadvantage that a delta comparison between two drawing versions (which is part of operational practice) is very difficult to implement and often yields poor results. This is different with 3D object models. Here, a comparison is made for each product attribute with precise results. The 3D model is also the starting point for numerical simulations to verify the physical properties of the product being designed. Other information to be stored includes, for example:

- Work plans,
- NC programs,
- Manufacturing aids,
- Inspection plans.

A digital twin of the real product is created, see Fig. 5.4. and compare with the previous chapter.

In the 3D world, the generation of mathematical models can be carried out much more easily. At the same time, it is also the starting point for process simulations, i.e., the installation of the part (such as a center console in a car) into the body. Here, it is crucial whether the installation can be carried out simply and without complications. Before installation, for example, the part/product must be manufactured (e.g., by a press). These production processes can also be simulated. In order for the processes in the factory to be built to run efficiently, so-called material flow simulations are also required, i.e., the

Fig. 5.4 Digital twin

simulation of all material movements in the factory. In addition to the parts to be installed and information about the final product, detailed information about the future factory is also required. This includes, for example, the geometric dimensions of the factory, of machines, the position of cable connections, supply lines, etc. In total, data volumes in the terabyte range are quickly generated. As robots and machines play an increasingly important role, the respective software programs must also be simulated, and the robot movements must also undergo a virtual check. The result is a complete digital image of the real factory—the digital factory, see also Fig. 5.5.

In order to use the digital factory not only for the construction of the real factory but also to control the real factory, further data is required—namely, the digital product memory. Here, all data from production is stored. Examples include measurement data from machines, production time, possible rework, etc. Using the target data from the planning phase, the actual situation can now be checked and controlled.

5.6.2 Consequences for Management

Digital factory projects are closely linked to the topic of vertical integration and at the same time form the basis and starting point for big data analyses and machine learning. As proficiency in dealing with the technologies described here increases, so does the economic yield. Through the digital product memory and digital product development, a closed digital chain is created, providing transparency. This is the starting point for product and process improvements.

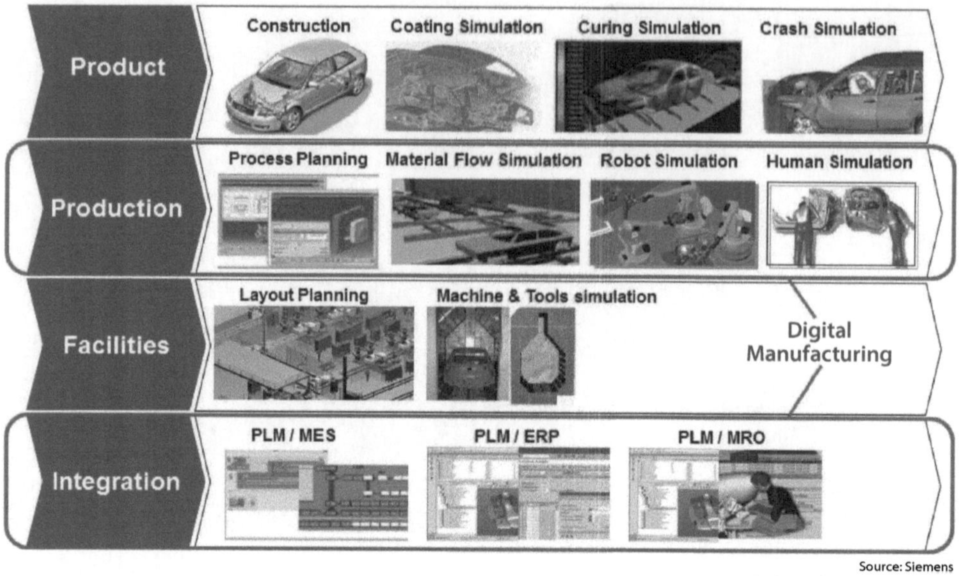

Fig. 5.5 Elements of a digital factory

5.6.3 Economic Relevance

There are many examples that have demonstrated the economic added value of digital factory approaches. In general, it is assumed that manufacturing costs can be reduced by about 5% as a result. At BMW, planning for changes to the assembly of a new exhaust system was reduced by 80%. In addition, travel times incurred could also be reduced by about 60%, see [20]. The American aerospace company SpaceX was able to increase its productivity by around 50% through the consistent use of digital product development approaches (using various Siemens products, such as Teamcenter as PLM and NX as CAD system), see [23].

5.7 Assistance Systems

The rapid development and widespread adoption of tablets and smartphones has fundamentally changed the way people interact with machines. The app principle—simple and intuitive operation of complex applications—is increasingly being adopted in the industrial sector. These applications deliberately conceal complex technical processes from the user. As a result, humans are moving to the center of human-machine interaction. Drivers of these rapid advances include new developments in sensor technology, actuators, machine learning, and real-time processing. Machines are increasingly becoming companions to humans and can adapt to their individual abilities. This applies both to the private sector (e.g., fitness bands or smart watches) and to industrial applications (e.g., sensitive robots). Such systems enable the combination of human abilities, such as creativity and empathy, with the capabilities of machines, such as freedom from fatigue and repeatability.

The range of technologies is very broad and includes:

- Smart glasses with Virtual Reality (VR) / Augmented Reality (AR),
- camera-based systems,
- RFID-based systems,
- mobile solutions,
- smart watches,
- pick-by-light systems.

With mobile solutions, the link to Real Time Enterprises is established once again; see Sect. 5.5. As has been mentioned several times, Industry 4.0 and digitalization are leading to strong networking of individual technologies. The broad spectrum is also illustrated in Fig. 5.6.

Areas of application for assistance systems can be found, among others, in:

- maintenance,
- employee management in production,
- logistics,
- training,
- development,
- quality,
- sales.

As visualization becomes increasingly important in big data applications, new solutions are also emerging here, for example, to allow users to "fly" through the data.

5.7.1 Technical Description

Due to the wide range of technologies, a compact presentation is extremely difficult. First, regarding terminology: AR (Augmented Reality) describes the visual overlay of the real world with additional information. Here, the information is displayed in the "correct" position and at the "correct" size within the field of view. Real-time capability of the interaction in all three dimensions is crucial, see Fig. 5.7; otherwise, delays occur between the real images in the user's field of view and the superimposed virtual information.

Fig. 5.6 Examples of assistance systems

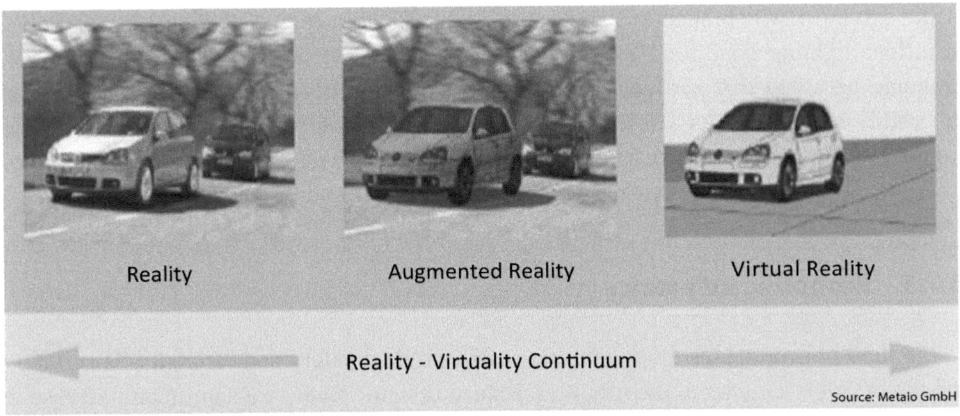

Fig. 5.7 Distinction between Augmented Reality and Virtual Reality

The goal is to provide the user with additional information in order to increase or improve productivity and quality of work. This results, among other things, in the following advantages:

- Cost reduction, as employees spend less time searching,
- resource optimization, by reducing machine downtime,
- quality improvement, through target/actual comparisons, contextual information, and work instructions.

In maintenance, mobile solutions have been in use for a long time. Another approach involves camera-based systems. These allow experts to remotely assess the situation and support their on-site colleagues in their work. An example of this can be found in Fig. 5.6.

Smart watches are increasingly finding their way into industrial everyday life. On the one hand, they serve as a replacement for mobile solutions (especially in maintenance, by displaying actions), but also in production. As is often the case, examples can be found in the automotive industry, such as at BMW. The cost of "simple" smart watches is less than €50. It does not necessarily have to be an Apple Watch. However, for the operational activities of maintenance personnel, for example as a replacement for smartphones, they have proven to be less practical.

5.7.2 Consequences for Management

Regardless of the technology, quite mundane issues often play a significant role in the acceptance of such approaches. The appearance and attractiveness of the devices should not be underestimated. Another acceptance issue (especially with AR and VR systems) is the extent to which such systems are visually flexible (e.g., for eyeglass wearers or in diffuse lighting conditions). Of course, price also plays an important role. However, it should be noted that hardware costs are continuously decreasing and miniaturization is steadily advancing. As a result, such devices will sooner or later find their way into logistics and production processes. Microsoft's smart glasses in the form of the HoloLens, costing over €3,500, certainly represent the "upper end of the known spectrum."

5.7.3 Economic Relevance

In the field of product development, increasing product safety and reliability is certainly the top priority. In general, the use of assistance systems leads to a significant increase in productivity in product development departments.

A similar picture emerges in production and logistics. The integration of real-time inspection into the entire production process leads to significantly safer workflows and reduces the required setup times and thus the order throughput time. Efforts for rework can also be reduced. Due to the high flexibility of assistance systems, rapid integration into existing processes is possible.

For example, DHL was able to achieve a 15% increase in productivity and a simultaneous reduction in error rates in logistics by using smart glasses. This is not, as is often the case, a first pilot project, but a large-scale deployment; see [2].

The benefits in employee training are often underestimated. Especially in connection with digital factory approaches, employees can be trained even before a factory is completed. Rare events that occur only occasionally can also be well trained in this way. This is where the connection to the digital factory takes place. The combination of VR approaches with the digital factory also supports employees in validating the concepts.

Such solutions are increasingly being used in sales, not only in the automotive industry but also, for example, in retail. The American department store chain Sakes is experimenting with such solutions for customers in the cosmetics sector; see [21].

The market volume for virtual assistance systems was around US$585 million in 2014. This value rose to around US$2.2 billion by 2019; see [9]. Overall, augmented reality can achieve productivity increases of between 8 and 15% in the relevant areas; see [12]. The AR & VR revenue in Germany will be around 1.8 billion euros in 2024.

5.8 Cloud

Cloud was already briefly mentioned in the last chapter in connection with platforms. For platforms, the technology is certainly a central, but not the decisive, element. Therefore, the topic will be revisited here and considered from a technical perspective. For this reason, it is not surprising that cloud technologies are often seen as a revolutionary approach. However, this often leads to a conflation between platform as a solution and cloud as the "underlying" technology.

The use of cloud computing refers to the demand-driven and flexible utilization of a wide range of IT services. These are offered over the Internet in the form of services. This enables companies to respond flexibly to new challenges and, if necessary, to open up new business areas through services and products, while simultaneously minimizing investment and ongoing costs. Building an IT infrastructure within the company is therefore not strictly necessary. The advantages of such an approach are manifold:

- optimal scalability of IT performance,
- high flexibility and demand-driven provisioning,
- greater degree of innovation through up-to-date software versions,
- improved availability and performance of the IT solutions used,
- focus on core tasks and relief of IT resources,
- cost savings,
- promotes IT standardization,
- faster and more straightforward rollout of IT solutions,
- increased IT security.

see also, among others, [22]. However, not all clouds are the same. The individual approaches are illustrated in Fig. 5.8.

Virtually all major IT service providers, such as IBM, SAP, or Microsoft, offer cloud-based solutions. The topic of cloud is not limited to "large" companies. Increasingly, SMEs (i.e., small and medium-sized enterprises) are using the cloud; in 2018, specifically, nearly 60% of companies with up to 2000 employees, which is very close to the figure for large enterprises at around 67%. Over the past years, acceptance of the approach has steadily increased, see [14]. By 2023, this share had risen to a total of 94% of companies using cloud services, see [31].

5.8.1 Technical Description

With regard to organizational form, a distinction is made between:

- Public,
- Private,
- Hybrid,
- Community

clouds, see Fig. 5.8. In a public cloud, services are made freely accessible by a provider and are thus available via the Internet. The exact counterpart to this is a private cloud. Here, the company itself provides the respective services or does so via an external service provider. The key point is that both data and functionalities are available exclusively to the company's own employees. The data may be stored either in the company's own

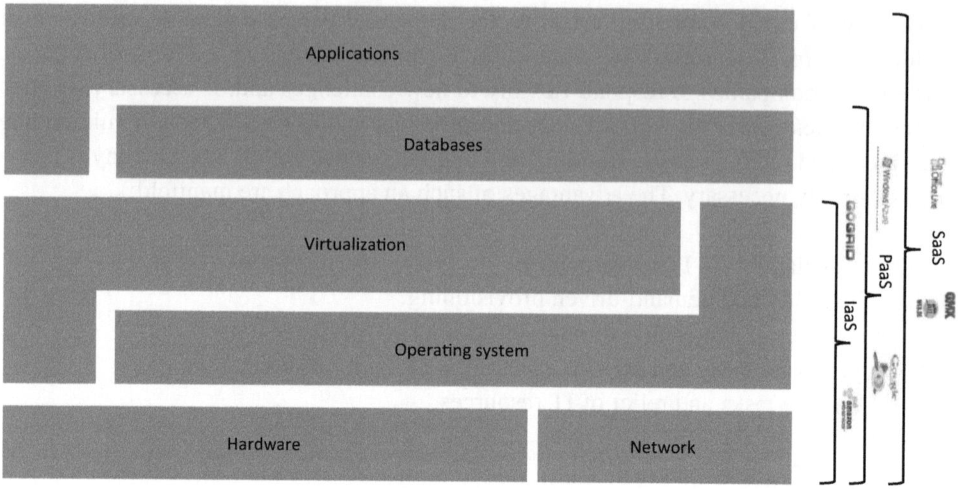

Fig. 5.8 Different cloud approaches

data center or in an external one. The data is transmitted via a secure connection. Furthermore, in the case of an external service provider, the location of the data center is known, which is not the case with a public cloud. The location is important for data protection purposes. Public clouds are mainly used for office software and non-critical data, see [14].

Most companies use a hybrid form between private and public cloud—that is, a hybrid cloud. A community cloud is often used in cross-company projects where project partners want or need to access shared data.

When selecting a cloud service provider, in addition to cost efficiency, the most important criteria are also outage and data security, data protection, scalability, and the availability of the offered services including their data (if they are to be stored in the cloud). Modern data centers are so-called twin-core data centers. In this setup, all data and systems are held redundantly in two data centers. This helps prevent possible outages (flooding, power failure, etc.). Data centers organized in this way are known as Tier 4 facilities. Data centers can be classified into four quality levels. Tier 1 has the lowest availability, while Tier 4 data centers are high-availability data centers with an availability of 99.999%, which equates to a maximum of just 5 minutes of downtime per year. Public cloud data centers are generally Tier 2 or 3 data centers. With Tier 2, the service is usually available immediately. Individual support from the provider is generally not provided. There is also a Europe-wide standard for the construction, operation, and management of data centers, DIN EN 50.600. In addition, the German Federal Office for Information Security (BSI) has published a catalog of criteria with over 100 requirements (Cloud Computing Compliance Controls Catalogue, or C5 for short). This gives cloud providers the opportunity to be certified by independent auditors, see [1, 4].

5.8.2 Consequences for Management

The currently greatest obstacle to more intensive use of cloud solutions is certainly the issue of trust in the security and reliability of the services offered, see [14]. The issue thus has both an emotional and a rational component. The emotional component can at least partially be addressed through high transparency (traceable key figures, localization of services in the German/European region), regulatory measures, and open communication. It is clear here that rational topics are used in an attempt to achieve an emotional effect. From a purely rational perspective, security is a technical challenge. In this respect, most people trust the cloud, see [14]. However, it should not be overlooked that, due to the very high electricity prices, data center capacities are growing much faster in other European countries than in Germany. For data centers, electricity costs account for about 40% of total operating costs, see [22]. The high and sometimes difficult-to-predict costs must be continuously monitored.

Regardless of which organizational form and configuration is chosen, it is important first to define the requirements for your own IT and cloud strategy. In many cases, companies cannot or do not want to afford their own IT infrastructure and personnel. This is especially true for smaller companies. In general, a hybrid cloud is significantly more complex than a public cloud. Unfortunately, a "turnkey" cloud solution is generally not available.

5.8.3 Economic Relevance

The Port of Hamburg has more than doubled its current goods and container movements compared to the year 2000. This was achieved through extensive measures, including the use of cloud technology. All logistical movements are recorded via this technology, and routes and parking occupancy are coordinated, see [20]. The EU Commission also sees significant economic turnover and potential in the topic of cloud computing, estimating approximately €250 billion in additional value creation and about 2.5 million new jobs in 2020, see [22].

Cloud technologies are also widely used among automobile manufacturers. Cloud approaches are part of the standard repertoire of IT departments in these companies. In many cases, these are hybrid solutions. The introduction of the "digital showroom" in London, for example, was able to increase sales by about 70%, see [26]. Cloud technology is a central component here. The widely valued connected services would also be unthinkable without cloud technology. AI-based approaches are giving the topic new momentum. The calculation of mathematical models requires significant computing power, which is best realized in the cloud.

The energy company BP Downstream has migrated its IT infrastructure and SAP applications to the public cloud, thereby reducing the costs for SAP operations by around 30%. The costs for the entire IT infrastructure were even reduced by 40%. BP is not an isolated case. As early as 2015, the Italian energy company Enel moved to the cloud. Their own data centers thus became partly a thing of the past. However, the motivation here was more strategic in nature: greater flexibility, scalability, and reliability were decisive for the cloud solution. Banks such as HSBC also use cloud services. In this case, the Google Cloud is used for data analysis. Another example among many is Boeing. The company uses Microsoft Azure cloud services for analytics applications, see [6]. These examples show that cloud approaches are a fixed component of many IT strategies, regardless of company size or industry.

However, the costs for storing and using data are not insignificant. This is leading to a certain rethinking. Primarily, only those data that are of global relevance or applications with high computing requirements should be stored in the cloud. The storage costs incurred in the cloud represent the proverbial tip of the iceberg. In addition, there are access costs for the data and processing costs. On the other hand, prices—and thus costs—for hardware and storage are continuously decreasing.

5.9 5G/6G

5G is the successor to 4G or LTE technology (Long-Term Evolution). Compared to LTE, the new standard offers several advantages, such as:

- Higher bandwidth, up to 20 gigabits
- Very high device density
- Very low latency of up to 1 ms,

see [3].

The standard was published at the end of 2018. According to Telekom, by the end of 2024, 99% of the population in Germany will have access to this standard. But after one standard comes the next. Work is currently underway on 6G. It is expected to enable data transmission speeds 100 times faster. By 2025, a corresponding network is to be established for the World Expo in Osaka, Japan. In addition, latency in data transmission is to be reduced to 100 ms (see [10]). The exact specifications of the new 6G standard have not yet been determined. Research and development have been ongoing worldwide since 2019.

5.9.1 Technical Description

Among the latest developments surrounding the 5G standard is its so-called "Advanced" extension. This enables the establishment of a positioning system. The topic itself is not entirely new. What is new is that this is now possible with a single technology. Until now, technologies such as RFID and Wi-Fi were required for this, see [8].

In general, the higher transmission rates are achieved through corresponding frequency ranges (3.4–3.8 GHz). This allows more data to be transmitted in the same amount of time. The disadvantage is the shorter range. Therefore, appropriate infrastructure is required.

5.9.2 Consequences for Management

5G is a key technology for autonomous driving. In the broader industrial context, however, the situation is somewhat different. The described advantages and features of 5G are insufficiently transparent. As a result, industrial companies find it difficult to develop corresponding use cases and to assess them financially. The technology must be evaluated in the context of Industry 4.0. In the field of production, the technology certainly plays a major role for German automobile manufacturers, but the situation is much more differentiated for their suppliers. Here, its use must be critically examined. Their plants are significantly smaller and less complex than those of the OEMs. The costs for 5G are continuously decreasing. Therefore, the economic effects must be continuously validated.

5.9.3 Economic Relevance

5G offers potential for many sectors of the economy. Examples can be found in logistics, agriculture, the gaming industry, and also for private households. Automobile manufacturers such as BMW use the technology in their plants (for example, in Leipzig). The economic potential in Germany alone is estimated at a mid double-digit billion euro amount. The establishment of so-called campus networks, for example for a production plant, is at the top of the priority list for large companies.

With the introduction of 6G, the economic potential will increase significantly and its industrial use will be greatly intensified. The initial euphoria surrounding 5G has given way to a more realistic assessment. It remains to be seen how developments around the upcoming new standard will unfold.

5.10 Metaverse

The Metaverse, also known as the Metaversum, is regarded in the tech scene and among its investors as the future of the internet and thus as the "next big thing." Accordingly, investments are being made in this technology. However, there is also increasing disillusionment surrounding this technology. Currently, the technology is still in its infancy. Nevertheless, there is a certain sense of optimism. In general, it represents the consistent further development of the internet, i.e., Web 3.0. There is currently no consensus regarding the terminology and the precise, substantive definition. This also explains why there is no unified vision of the Metaverse.

5.10.1 Technical Description

The technology is closely related to the already established blockchain, particularly in terms of commercial use. In general, the technology is in a growth phase and therefore exhibits many facets. This also applies to its innovation potential.

Regardless of the ongoing discussion, it is a form of virtual world or virtual experience. Thus, it is about a new form of interaction. Virtual worlds are already known from various computer games or from product development. There is also a seamless transition to simulations, such as those of factory buildings and the production processes taking place within them. This lack of clear distinction is also due to the aforementioned vague definition.

Technically, it is a decentralized approach and exists exclusively in the virtual world. Here, corresponding digital products exist (keywords: avatar and digital twin).

At its core, the business model is about the consistent further development of the platform economy. The aforementioned decentralized transactions and smart contracts in connection with blockchain and digital payments (bitcoins and their derivatives) enable its monetization.

5.10.2 Consequences for Management

The topic is currently overloaded with hype. Companies such as Microsoft, Meta, Siemens, and Nvidia are investing to turn the vision into reality. Therefore, it is important to critically monitor developments in this area. Current examples provide indications of the direction in which the topic could develop. Regulatory aspects are being addressed and, while they should be considered in implementations, they appear to be on a good path. Companies should be cautious about investing in specific technologies or solutions. This is also the advice from Gartner, see [30]. This is also reflected in recent developments at Microsoft. While the Metaverse was still "the" future topic for the company in 2021, just two years later, significant parts of the teams are being laid off and product development priorities are shifting towards AI, see [5].

5.10.3 Economic Relevance

There is no clear picture when it comes to assessing the potential economic opportunities. Since it is essentially about the reinvention of the internet, forecasts are sky-high. However, realistic assessments are difficult to make, and the forecasts are becoming more cautious year by year, with critics growing louder. Because the definition is vague, practical examples are also vague. Adidas and BMW provide initial examples. Possible applications include, for example, in the area of

- Work environment (new work with virtual meetings based on avatars)
- Education (virtual learning)
- Production and logistics (remote service or virtual commissioning)
- Healthcare (telemedicine)

and, as initial examples show, are already being implemented.

5.11 Quantum Computing

With quantum computing, societal challenges could be addressed in real time and previously unsolved problems could be tackled. Approaches from the fields of artificial intelligence, logistics, production planning, material analysis, and traffic planning are just a few examples, see [13]. The further development of quantum computing is a societal task. Despite significant progress in recent years, the technology is still in its early stages. Nevertheless, it is important to build up the necessary expertise in Germany and Europe and to drive development forward. The technology is attributed with great potential and influence on societal development.

Currently, high computing power is required for the use of AI solutions. For novel AI-based solutions, the computing demand increases significantly. Here, quantum computing (QC) could provide a solution. At present, the computing power of computers doubles every two years, while the effort required for AI-based solutions doubles every 3.5 months, see [13]. Thus, it is not surprising that industrial companies are looking for new approaches to meet their computing needs.

The first quantum computers, such as those from IBM or Google, are already available today. In general, the technology is very novel and at the same time financially demanding. This is certainly one of the reasons why this technology is currently primarily provided in the cloud and is thus easily accessible to users. BASF uses quantum computing for research, see [17]. BMW and Airbus use quantum systems, for example, in fuel cell research, see [18]. These examples show that this technology is slowly making its way into the business world.

5.11.1 Technical Description

QC uses quantum effects for information processing. While a classical digital computer works with bits and bytes, a QC operates with so-called qubits. These cannot be assigned to exactly one of two possible states; instead, a superposition of both is also possible. This allows for high computing power to be achieved with just a few qubits, enabling the processing of large amounts of data, see [13]. To this end, novel algorithms and mathematical models need to be developed. Quantum machine learning could open up new avenues in the field of AI.

5.11.2 Consequences for Management

Currently, the technology is still in its infancy. Nevertheless, it is important to monitor further developments in order to be able to react early. Building up in-house expertise is therefore not absolutely necessary at this stage. According to a 2022 survey by bitkom, 13% of the companies surveyed are planning to use QC or are discussing it, while another 2% are already using the technology in pilot projects, see [24].

5.11.3 Economic Relevance

The areas of application for QC are very diverse. Optimization problems top the wish list of users. Such challenges are found in virtually all industrial sectors, such as finance, procurement, logistics, or production. A common feature of these problems is the handling and processing of very large amounts of data and the analysis of various possible solutions. The optimization of traffic systems (keyword: traffic management system) or the aforementioned material analysis, as well as simulations of manufacturing processes and analyses, represent further areas of application. Thus, the economic potential is very high but also currently difficult to assess. In general, the potential market is estimated to be worth billions, see [28].

References

1. Bundesamt für Sicherheit in der Informationstechnik, Sicherheitsempfehlung für Cloud Computing Anbieter – Mindestanforderungen an die Informationssicherheit, 02/2012.
2. DHL setzt weltweit Datenbrillen ein, Lebensmittel Zeitung, Ausgabe 31, 4. August 2017, Seite 38.
3. Jürgen Fleischer, Albert Albers, Reiner Anderl, Jan Aurich, acatech IMPULS – 5G in der Industrie, 2021.
4. Moritz Förster, Cloud, aber sicher: C5 des BSI prüft Anbieter mit 114 Anforderungen, iX, 08.08.20017.
5. Preston Gralla, MS setzt auf KI und sagt Sayonara zu Metaverse, Computerwoche, 23.02.2023.
6. Wolfgang Herrmann, BP, HSBC, Boeing – Konzerne verlagern Kernsysteme in die Public-Cloud, CIO, 14.08.2017.
7. Masaaki Imai, KAIZEN – Der Schlüssel zum Erfolg im Wettbewerb, Econ-Verlag, 1986.
8. Jürgen Hill, BMW setzt im Werk Leipzig auf 5G, Computerwoche, 18.10.2022,
9. Henning Kagermann acatech (Hrsg.), Innovationspotentiale der Mensch-Maschinen-Interaktion, April 2016, http://www.acatech.de/fileadmin/user_upload/Baumstruktur_nach_Website/

Acatech/root/de/Publikationen/Stellungnahmen/acatech_IMPULS_Mensch-Maschine-Interaktion_WEB.pdf, aufgerufen am 08.08.2017.
10. Martin Kölling, Mehr Daten als 5G – Japan zeigt, was mit 6G möglich wird, Handelsblatt, 02.01.2023.
11. Sascha Maier, Sandra Aengenheyster, Geschäftsrisiko Cyxber Security: Leitfanden zur Etablierung eines resilienten Sicherheits-Ökosystems (essentials), Springer Verlag, 9. November 2020.
12. J. Manyika, M. Chui, P. Bisson, The Internet of Things: Mapping the value beyond the hype, McKinsey Global Institute, https://www.mckinsey.de/files/unlocking_the_potential_of_the_internet_of_things_full_report.pdf, aufgerufen am 23.08.2017.
13. Adrian Paschke, Nikolay Tcholtchev, Quantencomputing, Konrad-Adenauer-Stiftung e. V. 2021, Berlin.
14. Axel Pols, Marko Vogel, Cloud Monitor 2017: Eine Studie von Bitkom Research im Auftrag von KPMG, März 2017.
15. Jann Raveling, Wie der Milchbote: Der „dynamische Milkrun 4.0" beliefert Maschinen in der Fabrik nur bei Bedarf, Wirtschaftsförderung Bremen, https://wfb-bremen.de/de/page/stories/digitalisierung-industrie40/wie-der-milchbote-der-dynamische-milkrun-40-beliefert-maschinen-in-der-fabrik-nur-bei-bedarf, aufgerufen am 01.08.2017.
16. Guido Reinking, Die Cloud im Driver's Seat, in Ferri Abolhassan (Hrsg.), Was treibt die Digitalisierung, Springer Gabler, Dezember 2015.
17. Bernd Reder, Quantum Computing BASF nutzt Quantencomputing in der Forschung, CIO, 26.06.2023.
18. Berne Reder, BMW und Airbus nutzen Quantensysteme für Foschung, CIO, 18.08.2023.
19. Alec Ross, Die Wirtschaftswelt der Zukunft, Plassen Verlag, 2016.
20. Armin Roth, Einführung und Umsetzung von Industrie 4.0: Grundlagen, Vorgehensmodell und Use Cases aus der Praxis, Springer Gabler, Februar 2016.
21. Rod, Sakes zeigt Kosmetik per Augmented Reality, Lebensmittelzeitung, Nr. 24, 25. August 2017, Seite 34.
22. Dieter Schweer, Jan Christian Sahl, Die digitale Transformation der Industrie – wie Deutschland profitiert, in Ferri Abolhassan (Hrsg.), Was treibt die Digitalisierung, Springer Gabler, Dezember 2015.
23. Siemens Case Study, SpaceX delivers outer space at bargain rates, https://www.plm.automation.siemens.com/en/about_us/success/case_study.cfm?Component=30328&ComponentTemplate=1481, aufgerufen am 01.08.2017.
24. Tobias Schlichmeier, Bitkom-Umfrage zu Quantencomputing. Wirtschaftliche Relevanz steigt, elektroniknet.de
25. Andreas Streim, Simram Mann, bitkom Presseinformation, 31. August 2022,
26. Frank Strecker, Jörn Kellermann, Die Cloud in der Praxis, in Ferri Abolhassan (Hrsg.), Was treibt die Digitalisierung, Springer Gabler, Dezember 2015.
27. Günter Ullrich, Fahrerlose Transportsysteme: Eine Fibel – mit Praxisanwendungen – zur Technik – für die Planung, Springer Vieweg, 11/2013.
28. Steffen Wischmann, Unternehmen sollten früh Quantencomputing-Anwendungen entwickeln, Tagesspielgel, 13.04.2022
29. Stephan Wiesend, Polizei in Bayern und Österreich setzen auf iPhones, Computerwoche, 28.07.2017.
30. Jackie Willes, Gartner Online, Was ist ein Metaverse? Und sollten Sie sich beteiligen? 21.10.2022, www.gartner.de/de/artikel/was-ist-ein-metaverse, Aufgerufen am 25.02.2023.
31. https://sequafy.com/cloud-statistiken-trends, aufgerufen am 16.04.2023.

6 New Products Change the World and Companies

Networking via an always-on internet connection and through sensors/actuators is transforming products and, consequently, production itself. These new products are the Cyber Physical Systems (CPS) already introduced in Chap. 4. By providing new functionalities—primarily through software, smart sensors, and their networking—the functional potential is significantly increased. New business models are emerging based on the data generated by CPS.

As a result, traditional product boundaries, including those of the associated supply chain, are being broken, industry structures are changing, and entire sectors are being redefined. In the course of this transformation, the question of value creation naturally arises, see [7]. This concerns both one's own position and that of partners and suppliers. The shift in value creation inevitably leads to a changed competitive situation. Companies must face this change. In general, technologies are merely a means to an end, such as the internet in the Internet of Things (IoT), which is therefore a somewhat misleading term, since it is not the internet itself that is essential, but rather the functionality extended by software and the altered competitive landscape. Overall, the way work is done and business is conducted—from product development to sales—changes; the entire company must rise to the new challenges and adapt accordingly. Thus, business models also change, as do the ways and means by which companies earn their money.

In previous developments triggered by IT, products themselves remained largely unchanged. With CPS, entirely new products are defined through hardware/software, internet connectivity, sensors/actuators, and cloud technology—or at least, this possibility exists. Digitization also enables the transformation of real objects such as photos and videos into digital counterparts, allowing them to be completely altered and enabling the provision of tailored services (this is also referred to as the dematerialization of products). Digitization also brings a change in product providers. This shift can be illustrated by two examples: Brockhaus, after more than 200 years, stoped production

W. Huber, *Industry 4.0 Compact—How Technologies Are Transforming Our Economy and Our Companies*, https://doi.org/10.1007/978-3-658-51029-9_6

of its reference works in 2013. The company's product offering has shifted significantly toward digital services. In early 2015, RadioShack (an electronics company with sales of over US$4 billion) filed for bankruptcy with more than 2400 US stores, see [5]. Another example is Nokia. Until the launch of the iPhone in 2007, the company was the market leader in mobile phones from 1998 to 2011. The company failed to adapt to the new market conditions. Similar examples can be found in virtually every industry.

However, these changes also have positive aspects. They will bring about another significant leap in company productivity. This will also require all internal areas—from development and production to sales and aftersales—to change and adapt in order to realize potential opportunities and meet market and customer requirements.

The scale of this transformation is also evident in sheer numbers. Today, about 25–30 billion things are connected to the internet. In the coming years, this number will rise to over 50 billion. With a world population of around 8 billion people, this would mean that, on average, each person would have more than 6 connected devices, see [5]. As is often the case, these figures vary from source to source and from year to year.

One of the main drivers of this profound change is the rapidly evolving consumer behavior. Customers increasingly expect an "all-inclusive package" with ever more pronounced individualization and service features. Just imagine an iPhone without apps and an app store.

The technologies discussed in the previous chapters are now being integrated into a comprehensive economic concept. Here, products are paramount and come first. Without new business models, however, efficient production and new products would be worthless. Therefore, the next chapter will present a transformed production, followed by a chapter on new digital business models. But these lines of action are not sufficient—companies must also be aligned with the new conditions. This will be addressed in Chap. 9.

6.1 Transformed Products

Digitization and dematerialization are shifting the focus within product development. As a result, design and used materials become less important for such dematerialized products. Handmade paper (as in the above book example) does not exist in digital form, nor does a book's gilded edge. This also affects design and the entire product development process, which will be completely transformed due to changed (digital) product requirements. For example, digital connectivity in vehicles is a decisive purchasing criterion in China, while horsepower or panel gaps (i.e., the distances between individual components) are less important. These are just two examples of changing customer behavior.

The importance of product development is generally increasing significantly for companies, as is its complexity. In the future, products must be intelligent, i.e., smart—they will become CPS. We will now take a closer look at the key characteristics of CPS from a product perspective. These were already introduced from a technical perspective in previous chapters. Now, the focus is on the economic implications and consequences for companies. The characteristics are:

- expanded functionality through software and sensors,
- networking and communication,
- data,
- services,
- changed human-machine interaction.

However, a significantly expanded infrastructure is also required compared to conventional products, see Fig. 6.1 and [7].

These individual facets from the above list require closer examination.

6.1.1 Extended Functionality

The extended functionality in the form of software and sensors can be divided into different levels:

- Monitoring (of its own status as well as the environment), usually via sensors,
- Control (of its own functionality, personalization), for example via the cloud,
- Optimization (improvement of product performance and functionality) through additional software,
- Automation and autonomy (of the product's own operation),

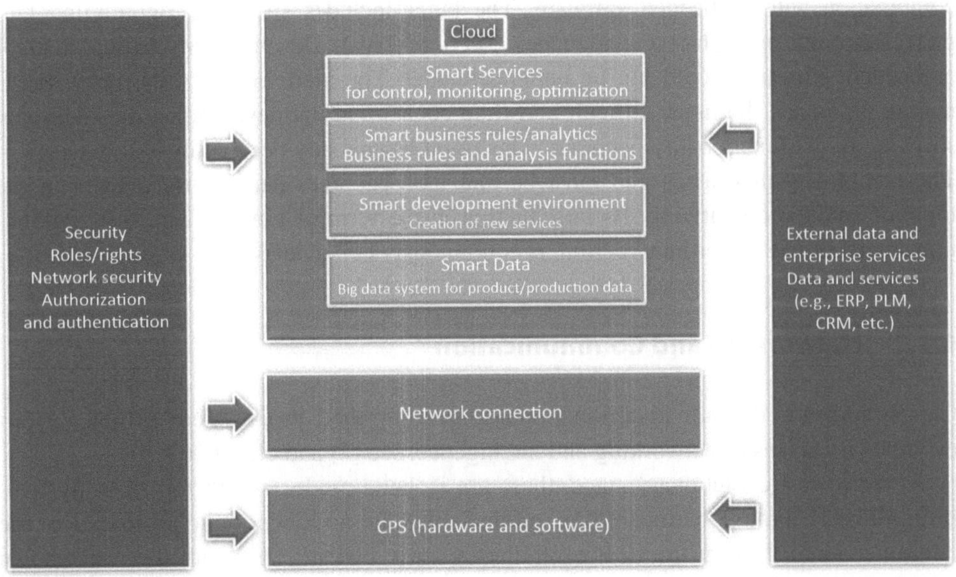

Fig. 6.1 CPS infrastructure

see [1]. The individual areas build upon each other. In the final stage, products thus encompass all four areas. The various characteristics can be found to varying degrees in lawn mowing and vacuum cleaning robots, as well as in sensitive robots and autonomous transport units or self-driving cars. Vacuum cleaners, for example, operate autonomously and clean the floor without human intervention. The prices for such vacuum robots are decreasing. The price here depends strongly on the degree of autonomy and thus on the built-in sensors and the associated software, i.e., how "smart" (i.e., autonomously acting) they are.

When monitoring machines and systems in production, for example, wear can be detected at an early stage before a product failure occurs. This applies to cars just as much as to manufacturing machines or wind turbines. Through optimization, wind turbines can be just as well adapted to wind conditions as autonomously controlled combine harvesters to their environment.

When controlling and optimizing new, smart products, additional functionality from the cloud can be accessed, for example in the form of apps. Especially in control, the functionality can also be statically embedded in the product. Access to the cloud is not always desirable. However, to use the full range of functions, the new products must be networked, including the appropriate infrastructure. On the other hand, with cloud-based approaches, corresponding updates can be implemented much more easily. In addition, the functionality of the physical product can be kept much simpler and more "timeless," since the majority of the functionality is implemented in the cloud.

The foundation for these new possibilities is provided by sensors and actuators in combination with embedded software. The latter includes specific programming languages as well as high-level languages (such as Java). Developments in the field of embedded software should not be underestimated. The more complex the "base product" (e.g., a car), the more difficult and time-consuming the development of additional software-based functionality becomes. This point has already been addressed in Chap. 4. The complexity that arises and must be managed becomes particularly evident in the topic of autonomous driving. Its introduction in the form of Level 5, i.e., fully autonomous without any driver intervention, is continuously being delayed.

6.1.2 Networking and Communication

Closely linked to the individual functional levels described, products can exhibit various evolutionary stages of networking and intelligence, see [7] and cf. Fig. 6.2.

At the lowest evolutionary stage, these are isolated products with functionality realized primarily through hardware (i.e., products that are no longer quite up-to-date in the consumer sector). From the next evolutionary stage onward, these are the well-known CPS in their various forms, i.e., products equipped with software and sensors that can

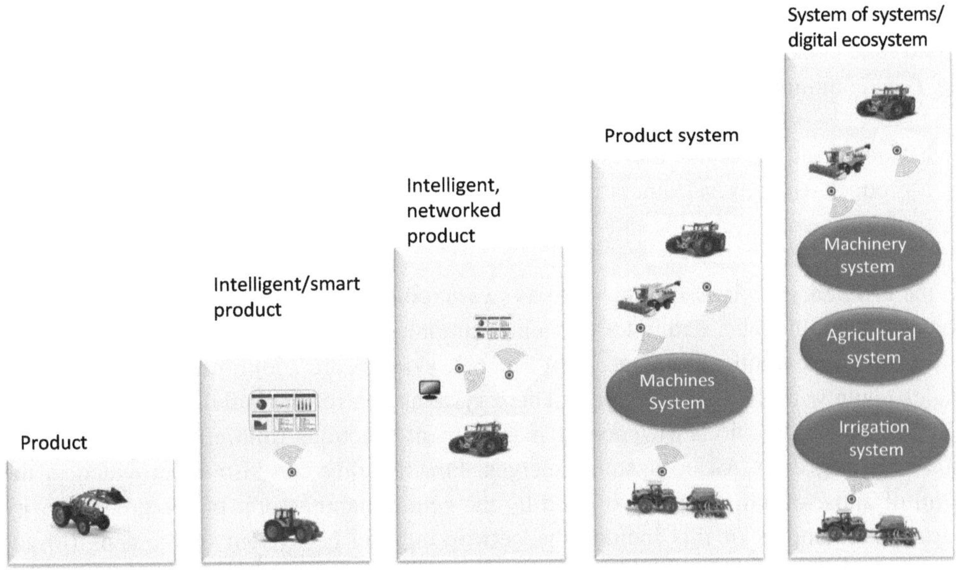

Fig. 6.2 Evolutionary stages of networked systems

perceive both the environment and their own status. CPS can also be networked. This enables both the exchange of data with others and the outsourcing of functionality, for example, to the cloud. Networking here refers both to networking with similar products and with others. Examples of networking with other products include engines in an aircraft that coordinate with the landing flaps, or tractors in agriculture that, controlled via satellites, distribute the optimal amount of seed. Networked elevators, for example, are able to increase transport capacity by up to 50%, see [7]. Here, networking occurs with other elevators, the hotel room locking system, or intelligent doors (as soon as a hotel guest leaves their room, they are highly likely to use the elevator. This scenario could be further refined using movement profiles of hotel guests).

The Kuka robots already mentioned in Chap. 4 are able, via the cloud, to provide additional functionality or, through data exchange, to optimize robot availability. Via the Internet and the cloud, updates (bug fixes, feature enhancements) can also be implemented without additional maintenance downtime (e.g., in workshops for cars). Networking also enables direct interaction between two products, such as two robots or a robot and an autonomously operating transport unit, to automate and thus optimize production processes—without human intervention. The autonomous transport units, in turn, coordinate independently regarding the execution of transport orders. Another example of networking between similar products would be car-to-car communication between two or more vehicles to warn of traffic jams or icy conditions.

Networked products are thus more powerful than conventional (purely hardware-based) and also than hardware- and software-based but isolated products. Thus, networking (i.e., communication) can occur between:

- a product and the cloud, the user, or another product (1:1 communication),
- a product with several others or many among themselves (1:n and n:m communication)

In the last item, this is also referred to as swarm communication, which is a prerequisite for swarm intelligence. Examples of such communications have already been mentioned.

In the next evolutionary stage from Fig. 6.2, systems are additionally involved (as in the example with the hotel elevator). These systems control the individual CPS and provide additional functionality, such as management functions (billing systems, accounting, statistics, etc.). An ecosystem emerges. Functionality can also be provided in the form of apps, which are either offered by the actual manufacturer of the product or by partners. Examples of this include the networking and ecosystem of Siemens (MindSphere and Xcelerator). At Siemens, for example, the individual S7 machine controllers communicate with the cloud and use a wide range of additional functionality. GE, through the analysis of data from airplanes and engines, has identified inefficiencies in flight times, routes, and fuel consumption of 18–22%. If only 1% of this could be realized, it would lead to savings of around 1–2 billion US dollars per year, see [9].

In the final stage, the system itself is embedded in a system landscape. The CPS then communicates with other systems, and the systems exchange information among themselves. To achieve differentiation, often only the apps of the respective manufacturer can communicate with each other. An example of a closed ecosystem is Apple. Their solutions work exclusively on the corresponding Apple infrastructure. The interoperability of the individual solutions with those outside their own ecosystem is unfortunately somewhat limited.

6.1.3 Data

The generation of data to create added value for one's own company, customers, partners, and suppliers is the central "pivot point." The goal is to master the "digital point of sale" or "digital touch point," in other words, the place where data is generated.

Networking is not a one-way street. Intelligent products generate a large amount of data. This data can be specifically used for further optimizations and business ideas. This feedback of data is therefore of considerable importance. In general, companies must build a completely new infrastructure, see Fig. 6.1, see also [7]. This is only superficially a technical problem. Building an infrastructure as shown in Fig. 6.1 is, from a purely technical perspective, certainly a challenge for companies. However, implementing this into new products and profitable business models is a much greater challenge. This involves both internal company challenges as well as those posed by the market. The market must accept the new products.

Smart products must be able to communicate with the company's own infrastructure. In advance, it is necessary to define which data is relevant (for the company itself, customers, partners, and suppliers). Since smart products can also be machines in the customer's production, customer-specific needs must also be taken into account. Security is of central importance here (for example, encryption of communication or communication via secure channels, etc.). On the infrastructure side, big data systems must be built, both for the products and for further business models. Data in a wide variety of formats and from different sources flows into the big data system (temperature, air pressure, machine status, etc.). It is also necessary to use external data sources and functionalities or to establish mechanisms for their secure integration. This also concerns the integration of the company's own systems. The solution approach is the vertical and horizontal integration already discussed in the previous chapter.

Technologically, it is essentially necessary to master all the topics and technologies addressed in Chaps. 4 and 5. The use of data is therefore a very central but at the same time highly complex requirement for companies. The technological breadth of the topics suggests that it will not be easy for companies, regardless of their size, to build up this know-how from an economic perspective. This applies in particular to SMEs. An example of this complexity is the digitalization of automobiles. Here, individual manufacturers are making great efforts to position themselves, but so are tech companies such as Apple, Google, or Amazon.

In data generation, the new smart products in the form of CPS are certainly a starting point. On the other hand, customer needs must be extracted from social media and other data sources. Other sources include the company's own employees, customers, partners, and suppliers.

6.1.4 Digital Services and Platforms

In general, companies must gradually (i.e., in the medium term) move away from the sole production of individual products. Customers expect an all-round service, which is associated with the shift towards becoming a service provider. The goal is to offer additional functionality or services. However, caution is advised against overly general statements. Especially for very durable or highly complex products, the classic approach of selling may still be suitable and even in the long term. Nevertheless, additional services must also be developed here. Performance-based contracts could establish themselves as a "golden mean."

Regardless of the approach, this transformation is not unproblematic for companies. It means nothing less than a cultural change. This change affects all areas of the company. But it gets even more challenging: this transformation cannot be accomplished without software and data in the form of new services. Thus, companies are facing a double cultural shift. Automotive manufacturers have already partially undergone this break, at least regarding the importance of software. A large part of product innovations in this industry are realized through software. Software-based innovations are associated with the advantage of corresponding weight savings and, in many cases, cost benefits (zero marginal costs).

As a logical consequence of smart services, the entire product development process will be completely different in the future—and in many cases, it already is. Software developers will have at least the same status in and during product development as design engineers. Both sides must work closely together. The E/E development (electrical/electronic) in automotive engineering could serve as a model here. The combination of hardware and software has long been practiced in this field. This approach, in turn, has been strongly influenced by "classical" software development.

One advantage of this changed product development based on software is that product variability and thus adaptation to specific markets can be achieved much more easily via software than solely through hardware. However, this variability must be planned for and taken into account during product development. Specific country adaptations can then be made later during the "normal" product life cycle.

As a prerequisite, smart products that offer additional services must, of course, collect the relevant data. For this, sensors and interfaces must be defined during the development phase. Copier manufacturers, for example, have long since evolved into providers of copying services. These providers no longer sell copiers, but rather a service based on actual usage—measured by sensors and controlled by software. This shift towards becoming a service provider also impacts product design. Among other things, it is necessary to implement service interfaces and critically verify the durability of components, since the focus is no longer on the product itself, but on the service.

The new smart products must be continuously developed further. The continuous improvement process (CIP), known from the lean world, is therefore also important in the digital world and serves as a tool for customer retention and as a central management instrument. This enables added value for the customer, as the "aging process" of the product is changed. Continuous product improvement and adaptation to new circumstances is possible via software, see [8].

The security aspect already addressed and outlined in Chap. 4 and in Fig. 6.1 must not be neglected under any circumstances. This is also a way to build trust with customers in the digital world (regardless of whether they are end customers or partners). IT security can even become a value driver and a differentiating feature. However, this must already be taken into account during the development phase. IT protection implemented retrospectively can generally be circumvented much more easily.

Regardless of the efforts companies must make towards smart products and corresponding services, this functionality must be offered to the customer via an appropriate infrastructure. This is where platforms come into play again (see Fig. 6.1). Platforms have several facets. On the one hand, they are essential for the distribution of one's own services. Examples include the AXOOM platform (founded by Trumpf) or Apple's App

Store, as well as established automation solution providers such as Siemens. However, platforms also offer the possibility to "insert themselves" between the product manufacturer and the end customer, as Amazon or Zalando have done very successfully. Platforms also support companies in the use of technologies. Through the platforms of IBM or Microsoft, it is possible to use their solutions and partner solutions (see Chap. 4). However, the examples of AXOOM and GE also show their pitfalls. For instance, Trumpf sold its platform to GFT in 2019. GE's cloud solution has also fallen far short of management's expectations.

6.1.5 Human-Machine Interaction

An important aspect of new products is human-machine interaction (HMI). The topic is technologically broad. It brings together a large portion of the technologies already discussed in Chaps. 4 and 5, such as machine learning and vision (i.e., image recognition), sensor technology and actuators, real-time capability, assistance systems, vertical and horizontal integration. It is therefore crucial to link the individual technologies together in an "intelligent" way.

The goal and content is the intuitive operation and use of technical systems with little or no prior knowledge. This enables an almost natural interaction. In 2018, around 20% of Google search queries were made via voice. In 2020, this figure is about 50% and rising, see [2]. Voice input has become established in the consumer sector since Apple's Siri, Amazon's Alexa, or Google's Home. In the professional environment, IBM is pursuing a similar approach to interaction with Watson. AI-based approaches such as ChatGPT clearly indicate the direction and potential. Gesture control is also increasingly being used, especially in robotics and automotive operation. In robotics, gesture control serves as a substitute for programming and can therefore be performed by a much broader user base. Specialized knowledge in the form of programming skills is thus required to a much lesser extent. In the field of mobility, assistance systems also support older drivers, which will be further advanced by autonomous driving and intelligent navigation systems. Automated passenger and freight transport can, for example, reduce the number of traffic jams.

Apps are a positive example of easy operation, also from an economic perspective. In Europe, the app economy provided around 1.64 million jobs in 2016 (which corresponds to about 0.7% of all jobs), see [3].

Worldwide, double-digit growth rates are forecast for HMI technologies. This will also have an impact on value chains and industry boundaries. German companies in particular have considerable catching up to do in this area. There is great potential, especially in the healthcare sector. This sector contributes around 10% to Germany's gross value added. Evidence-based expert systems are already being used today to support diagnosis and therapy. These systems draw on the knowledge of thousands of clinical studies. The global digital health market amounted to around US$135.8 billion in 2017 (an increase of 21% compared to 2013). In 2020, the volume rose to around US$233.4 billion. According to Statista, the global market will exceed one trillion US dollars by 2025, see [10]. Another facet is the shortage of skilled workers. Digital solutions can automate or replace individual activities with smart solutions (such as service robots). This makes it a lucrative market for optimizations in the form of smart solutions. Similar figures apply to assisted driving. Here, sales amounted to US$4.38 billion in 2014. In 2020, this value rose to around US$17 billion. German companies (manufacturers and their suppliers) are world market leaders in this technology with a market share of around 52.5%, see [4]. The use of innovative HMI approaches often raises legal and liability issues, as well as data security and data protection concerns. These must be addressed and clarified at an early stage. One example is the EU's GDPR regulation.

6.1.6 Transformation: Development and Definition of New Functionalities

When selecting the "right" products and functionalities for a given company, an appropriate strategy is required. The resulting questions are outlined in Fig. 6.3, see also [7].

When choosing new functions, it is important to keep an eye on customer benefit versus the costs incurred. This customer focus has been well known since the Lean era. In addition, it is also necessary to work out differentiating features. So, in summary, these are not really new questions. Even in the new digital world, it remains true that the customer must be satisfied and their requirements met in order to remain competitive.

In implementation, many companies will certainly handle this partly in-house and also selectively or at the start of this new world rely on external partners. This hybrid approach makes perfect sense, as the required specialists for dedicated topics are not

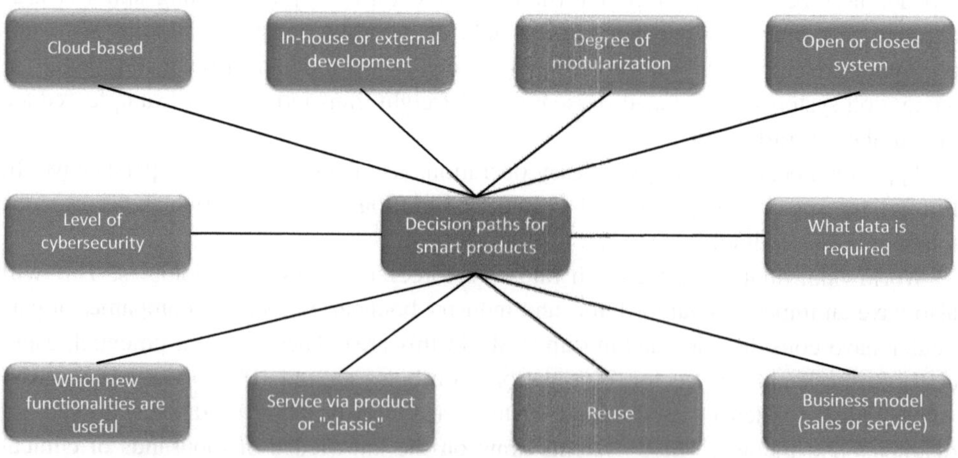

Fig. 6.3 Key questions in the design of smart products

necessarily available and it may not be sensible to employ them permanently in the company. This approach has already proven itself many times in practice, even in large companies.

The degree of modularization is also important, i.e., the creation of standardized and combinable functional units including aggregation into customer-specific (service) products. This allows the potential of classic custom development to be combined to some extent with that of standard software development, enabling individual products to be presented to customers quickly. Similar approaches are already used in classic product development. The same applies to reuse. This "old dream" of computer science should not be missing here. Both points mentioned have a decisive influence on the profitability of individual functions and the entire product.

Since smart products can recognize and check their own operating status, the question arises as to whether a classic service that checks the product or machine status at fixed intervals is still necessary. This can certainly be provided much better by the product itself. Thus, service can at least be partially digitalized. However, necessary repairs still have to be carried out by humans.

In general, smart products can deliver the value proposition to the customer much more easily and effectively than classic products. It is therefore not surprising that former GE CEO Immelt stated in March 2014 that every industrial company will sooner or later become a software company, see [7].

6.2 Novel Product Development Process

All the points listed so far lead to a novel product development process. This is also necessary in order not to fall behind amid the rapid pace of technological progress. Development must generally be both software- and hardware-based. It should, as a rule, be agile. Thus, development proceeds in small, incremental steps. To obtain customer feedback as quickly as possible, it is advisable to first develop a basic functionality that already delivers significant added value to the customer. The term "Minimum Viable Product" is often used for this approach. The product can then be further developed step by step. If customer feedback is negative, the development activities should be reconsidered. However, it is not advisable to leave further development solely to customer feedback. Especially disruptive products carry the risk of initially finding little favor with customers. Regardless of customer feedback and its importance in product development, an agile approach significantly shortens the time to market for new products, and does so at considerably lower costs (at least for the first version). These approaches, in sum, are not truly new. In practice, they are increasingly being applied.

This also requires a change in culture. Failure should not be viewed solely in a negative light. It is part of the overarching product development approach.

Customer feedback can be obtained either actively by soliciting it or passively by collecting data generated during the product's lifecycle (the latter is significantly more time-consuming and can therefore only be incorporated into a long-term product strategy). For evaluation and analysis, this data should be collected in big data systems. This opens up entirely new perspectives on the market and end customers. To access this data, a communications infrastructure is first required to read out the data. Cloud technologies are certainly the key to success here, see also Fig. 6.1.

Here, too, the automotive manufacturers are leading the way, including the internal problems and, in some cases, turf wars between proponents of hardware-based products and those of smart products. The industry is also increasingly recognizing the central importance of data. In the future, engine performance will play an ever smaller role in the decision to purchase a vehicle; instead, the digital ecosystem of the respective manufacturer and its ability to offer customers data-driven value-added services will be decisive. This trend is already visible in China, as discussed in previous chapters.

Open innovation and open development can not only reduce costs, but also open the company to creative and novel ideas and approaches. This, too, requires a change in corporate culture.

The development process itself must be continuously improved. The goal is to make meaningful use of the data obtained during product development for further product development. In this context, the term Data Driven Development (D3) is also used. In this approach, virtual validation is used not only to digitize product development, but also to increase its efficiency, starting as early as the initial development phase. Issues of reuse and automated testing, both within the system being developed and with regard to its external communication behavior, must be appropriately verified. The test results can then be used, via big data analyses, to draw conclusions about the actual maturity level of product development. For example, when testing an infotainment system in the automotive industry, a three-digit gigabyte volume of data can be generated per day. Thus, the use of big data systems is entirely justified and sensible. Generative design goes a step further. Here, computer-aided design (CAD), virtual reality, and artificial intelligence are combined. Development engineers define the framework conditions, and the AI generates solution proposals. This, too, can shorten development cycles, see [6]. The digital twin is also playing an increasingly important role here.

Development engineers must also adapt due to 3D printing. Depending on the chosen approach, additive manufacturing opens up new possibilities for product development. Bionic design, i.e., drawing inspiration from nature, becomes much easier to realize. However, product engineers must learn a new way of thinking for this. In general, irregular, asymmetrical, and complex structures are best suited for additive manufacturing.

6.3 Changed Competitive Situation

The disruptive technologies already discussed in Chap. 4 have the unpleasant characteristic of completely disrupting established development processes. Established companies should not make the mistake of allowing competitors (especially start-ups) to enter their own industry through their own inactivity. It is therefore better to cannibalize oneself than to leave the field to the competition. The German car manufacturers have been doing this extremely successfully for years. By constantly introducing new product variants, they prevent competitors from entering their market segment. Companies must recognize the opportunities of new technologies at an early stage or, better yet, develop these new technologies themselves.

Smart products make it possible to build a completely restructured ecosystem. Based on platforms and cloud-based approaches, not only can the company itself provide new functionalities, but partners can also be integrated. The quality of the additional functionalities (usually in the form of apps) and the reputation of the partners largely define—and hopefully enhance—the value of the company's own product. This also helps to further build switching barriers.

The market entry of competitors can also be prevented by offering more complex products, as this requires overcoming significantly higher financial and technological entry barriers. For example, a company in the agricultural sector can not only manufacture tractors and combine harvesters, but also provide a complete agricultural management system with a seed database, irrigation information, soil sensors, and specialized management software—across brands. This is currently being practiced by companies such as John Deere or Claas.

In general, connected products are more difficult—or only with great effort—to displace or replace. One reason for this is that they can be easily adapted to the individual needs of customers.

On the other hand, market newcomers can attack established manufacturers by skillfully exploiting new technologies and thus products. In the agricultural sector, independent providers of agricultural management systems have emerged, such as FarmFacts in Germany and OnForm in the USA. Both connect machines from a wide variety of manufacturers and integrate them into their overall solution. In the smart home systems sector, traditional manufacturers such as Bosch are facing serious competition from Apple. A similar situation arose a few years ago with Tesla in the automotive sector. Through software-based functionalities, the company suddenly became a strong competitor to established car manufacturers and was able to differentiate itself—first in the segment of a Daimler E- or S-Class or a BMW 7 Series, and now with the Model 3 in the segment of a Daimler C-Class or a BMW 3 Series.

The HMI technologies already mentioned also pose a significant threat in terms of user experience. These technologies, discussed in Sect. 6.1 and in Chap. 5, can also be used to enhance the customer experience. Numerous application examples have already been discussed. The "intelligent" linking and use of individual technologies is of considerable importance for the future positioning of companies, as well as for Germany as a business and scientific location.

However, the downside must also be considered. The negative examples of failed approaches already mentioned in this chapter show that this is a very complex and, unfortunately, both time-consuming and financially costly undertaking.

References

1. Birgit Buchholz, Jan-Peter Ferdinand, Jan-Heinrich Gieschen, Uwe Seidel, Digitalisierung industrieller Wertschöpfung- Transformationsansätze für KMU, Begleitforschung AUTONOMIK für Industrie 4.0, April 2017.
2. Peter Gentsch, Von Bot zu Bot, Lebensmittel Zeitung, NR. 31, 4. August 2017, Seite 38.
3. Kathrin Hollmer, Gute Arbeit, Aufbruch Erfinden, Google Inc., Anzeigenveröffentlichung, 2017, Seite 4–7.
4. Henning Kagermann acatech (Hrsg.), Innovationspotentiale der Mensch-Maschinen-Interaktion, April 2016. http://www.acatech.de/fileadmin/user_upload/Baumstruktur_nach_Website/Acatech/root/de/Publikationen/Stellungnahmen/acatech_IMPULS_Mensch-Maschine-Interaktion_WEB.pdf. Aufgerufen am 08.08.2017.
5. Ralf T. Kreutzer, Treiber und Hintergründe der digitalen Transformation in D. Schallmo (Hrsg.), Digitale Transformation von Geschäftsmodellen, Springer-Verlag, 2017, Seite 33–58.
6. Chris Löwer, Auf Datenreise, AutomotiveIT, 2/2023.
7. Michael E. Porter, James E. Heppelmann, Wie Smarte Produkte den Wettbewerb verändern, in Harvard Business Manager, Dezember 2014, Seite 34–60.
8. Michael E. Porter, James E. Heppelmann, Wie Smarte Produkte Unternehmen verändern, in Harvard Business Manager, Dezember 2015, Seite 52–73.
9. Daniel Schallmo, Andreas Rusnjak, Roadmap zur Digitalen Transformation von Geschäftsmodellen, in Daniel Schallmo (Hrsg.) Digitale Transformation von Geschäftsmodellen, Springer Gabler, November 2017.
10. https://de.statista.com/statistik/daten/studie/1179454/umfrage/digital-health-marktvolumen-weltweit-nach-segment/ aufgerufen am 21.5.2023.

7 The Smart Way of Production

The previous Chaps. 4 and 5 have already described the individual technologies that, among other things, are giving today's production a new face and will do so even more in the future. Automation and networking have already changed production and will continue to do so. We are currently only witnessing the beginnings of this profound transformation. Paper-based processes will finally become a thing of the past—and must do so, otherwise economical production, especially of batch size 1, will be difficult to achieve. In their place, transparent, real-time controllable, and paperless processes are emerging. Here, real-time means having the right data at the right time, in the right place, and at the right level of granularity. A plant manager is interested in aggregated and thus different information than, for example, a shift supervisor. The same applies to the temporal granularity of data and information. The increased transparency thus leads to greater flexibility in production.

Many processes, for example, are handled more efficiently and with a higher degree of automation and autonomy than is currently the case, thanks to sensitive robots and autonomously operating transport units. The factory of tomorrow will by no means be devoid of people, even though this is often claimed. However, people will take on significantly more supervisory tasks and less "hands-on" work. Thus, following the lean transformation, another transformation is taking place—this time a digital one. The efficiency gains of a smart factory are likely to be in a similar range as those achieved during the lean transformation—namely, in the range of 20–30%, see [2]. These are not merely marketing statements from product manufacturers or consulting firms, but rather concrete practical results, see [8].

These optimizations generated in production result from several areas in total:

- increased machine utilization,
- improved productivity of the respective employees,

W. Huber, *Industry 4.0 Compact—How Technologies Are Transforming Our Economy and Our Companies*, https://doi.org/10.1007/978-3-658-51029-9_7

- lower personnel costs,
- reduced quality costs and thus reduction of scrap and rework,
- lower administrative and planning effort,
- reduced resource consumption.

However, the smart factory will also have an impact on logistics. There is still considerable potential for automation and thus optimization, especially in the warehouse area. The same applies to the transport of parts in production and maintenance.

To implement this "brave new world," appropriate integrated systems and technologies are required. In contrast to business models, the transformation in the factory will, however, proceed in an evolutionary manner. Here, new technologies must be introduced step by step while preserving the advantages of the lean transformation. The change can be represented in three dimensions:

- manufactured products,
- machines and equipment used,
- production processes and production system.

Drivers of these changes are new products, but the production of "conventional" products will also change sustainably. Ultimately, technological change leads to completely transformed production systems and the entire value chain.

7.1 Smart Factory

A smart factory is the logical continuation of lean production approaches. It involves optimizing the entire customer order process—or, in modern terms, the order fulfillment process—that is, all activities from order entry to delivery of the product to the customer, regardless of whether the customer is an end consumer in the B2C sector or a company in the B2B environment. The implementation of a smart factory exists in various maturity levels and specific forms, see [8]. Currently, the most prominent example from automotive production is Daimler's so-called Factory56, see [9]. Naturally, the specific form is highly dependent on the industry. As the individual technologies continue to evolve, so too will the target vision of a smart factory. In many cases, individual scenarios and concrete use cases are implemented. Regardless, the required elements can certainly be identified. These are, as already mentioned, the technologies described in Chaps. 4 and 5, see also Fig. 7.1. Therefore, no further explanations are needed here. Of course, the technologies to be introduced depend on the specific context. Companies should only implement what makes economic sense. As is often the case, networking and communication between the individual technologies are crucial—more on this shortly.

One example is mobile solutions. Especially in the area of maintenance, there are significant opportunities to increase productivity. Maintenance personnel spend a

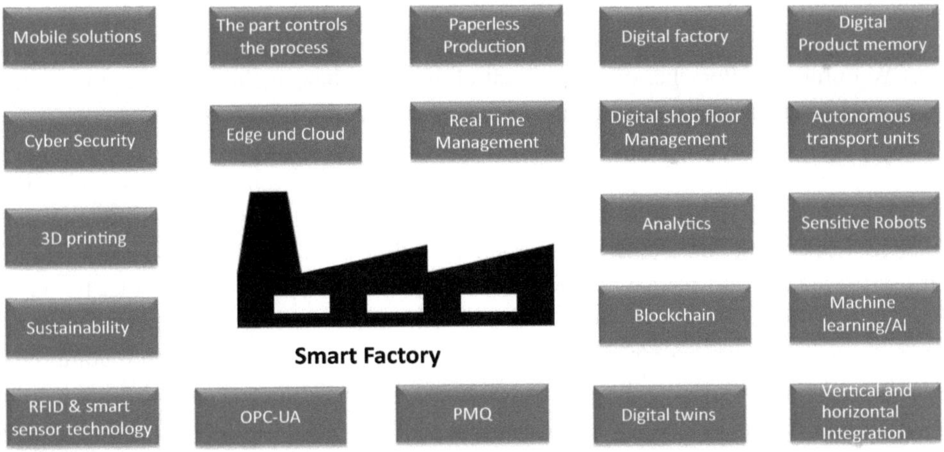

Fig. 7.1 Smart Factory

considerable portion of their work traveling between different areas, particularly in paper-based processes. Here, mobile solutions can deliver significant productivity gains in combination with improved data quality.

For all the discussion about technology, it is essential that these technologies are interchangeable in a smart factory in the future. Individual technologies simply change too quickly to warrant a prominent position in a factory. To achieve this, it is important to rely on appropriate standards. Furthermore, it is important that every workstation (regardless of whether it is manual or machine-based) can be addressed via an IP address and thus configured. Another sobering point is that it is not the technology, but the business process and its optimization, that should be the focus. There is little point in making inefficient processes appear more efficient through technology.

7.1.1 The Part Controls the Process

Regardless of the descriptions in the preceding chapters, some points that have not yet been mentioned will be addressed here. Once again, the focus is on the much-discussed networking. Even in production, the combination of individual components creates additional added value, which is absolutely crucial.

"The part controls the process" is a central element in the smart factory, especially when there is high product variance (for large batch sizes, such approaches make little sense; see also the above remark on industry-specific characteristics, see [5]). In this context, the logic that was previously stored mainly in a central MES is shifted to decentralized process elements, such as autonomous transport units, intelligent parts or machines, and sensitive robots. This breaks up rigid, centralized production structures (keyword: ISA95 model). The production islands known from the world of lean production can be seamlessly integrated into this concept, as demonstrated by examples such as SEW-Eurodrive, Porsche, or Audi. Central elements here include RFID, among others. Barcodes are no longer sufficient in the smart factory. RFID enables, among other things, active communication between intelligent parts/products and the MES, machines, and autonomous transport units. Another element is machine-to-machine communication (M2M). The part "knows" its status and, according to the production progress and resource availability, can trigger further activities. For example, the product notifies the transport system of a transport requirement, and the system can then decide, based on availability, which autonomous transport unit will take on the task. Whether this transport system operates on a swarm basis or is a shop floor-specific software solution is initially irrelevant. Here, too, the end justifies the means. The machines also know their own (production) status, including the pending production orders, and can communicate accordingly with other process participants via M2M. The MES stores and provides the relevant process data.

This decentralized control does not contradict central production planning. This remains necessary for the time being. Customer delivery dates must be met. Thus, central planning sets the "guardrails" for production, including material supply. Decentralized control, on the other hand, benefits the shop floor supervisors. In production, more or less pronounced turbulences or disruptions in the production flow often occur. These can be intercepted and corrected by this decentralized approach. For humans, the very demanding task arises of monitoring and controlling this swarm of intelligent elements.

The introduction of this approach, which at first glance appears very revolutionary, is actually quite evolutionary. Once again, the starting point is stable and efficient (i.e., lean) processes. A possible implementation scenario starts with RFID, which lays the infrastructural foundation. Initial benefits can be realized, for example, through significantly more efficient container and tool management or by avoiding scanning processes. Logistics can also benefit from this—of course, depending on the initial situation. In parallel, when purchasing new machines, attention should be paid to connectivity via OPC-UA. This is possible with virtually all manufacturers. This allows new machines to be integrated into the production environment more easily and quickly. It is important to ensure that the underlying control code (PLC code) includes appropriate mapping to OPC-UA. Decentralized control is not yet required at this stage. However, the costs are not insignificant. When introducing RFID, expenses in the range of five to seven figures in euros should be anticipated, depending on the number and size of production lines and warehouses. Therefore, a concrete cost-benefit analysis is essential at this stage. The end scenario must be kept in mind. In a further step, autonomous transport units and sensitive robots are introduced. The specific implementation scenario primarily depends on the actual need. Autonomous transport units are often introduced first, as their cost-effectiveness is very easy to calculate. The same applies to automation solutions and the associated use of sensitive robots. Subsequently, individual production lines can be converted to decentralized control. It is advisable to start with a "simple" line. In general, when introducing new technologies, it is best to pick the “low hanging fruit” first. Even in the smart factory, there is no clear recipe for success. It should be noted that, in general, digital solutions involve lower investment costs than, for example, automation solutions.

Regardless of the specific implementation scenario, central MES do not lose their significance. Rather, there is a functional relief and a gradual migration toward a CPPS (Cyber Physical Production System) and its continuous transformation.

The approaches of decentralized control are being consistently advanced in the automotive industry. They are also necessary in the sector to cope with the steadily increasing complexity. Driverless robotic vehicles handle the transport of car bodies through assembly. The systems independently decide which production island to approach next, depending on availability. Thus, the assembly line, at least from a future perspective, is becoming obsolete. Audi refers to this as modular production. The goal is to increase efficiency by around 20% compared to classic assembly line production, see [5]. To drive development forward, Audi has, for example, invested in the start-up Arculus. Further examples can be found at BMW, Porsche, or Daimler in various forms. Initially, the individual technologies were experimented with to find their own path. The comprehensive rollout is now the next step.

Regardless of whether a central or the decentralized approach described above is chosen, data will remain of fundamental importance in the production of tomorrow. A unified data basis is extremely important here. It supports process continuity. However, the urge to collect data should be curbed right from the start. Here, too, unnecessarily stored data represents a form of waste, which should be avoided (even in times of digitalization). Lean principles apply once again. Only data relevant to production control and its monitoring should be stored in the MES. In addition, traceability data is stored here to provide evidence of correct manufacturing.

The situation is different when it comes to big data. Here, it is important to ensure that process data are stored in a separate infrastructure from the MES. Such data has no place in an MES or CPPS to that extent, as it involves mass data. The relevant data will change over time due to the learning effect and the associated production optimization. The storage duration also varies greatly. Traceability data must be stored for decades, whereas process and machine data have a lifespan of about a year. In all the enthusiasm, it is essential to ensure an appropriately designed IT infrastructure in the form of sufficient transmission and storage capacities. Edge- and cloud-based approaches are used here. The infrastructure must also be set up for mobile solutions. Therefore, comprehensive WLAN coverage is required.

Big data also helps to optimize the production sequence. Based on historical data (production planning and actual sequence), further optimization potentials for the future can be derived. They support central planning and/or provide the local swarm of intelligent machines with appropriate input.

7.1.2 Robotics

Robots will increasingly take on more tasks in production. Using the capabilities introduced in Chaps. 4 and 5, these smart assistants can also take on tasks in the field of quality inspection. In addition to increasing automation, there is also the advantage of continuous documentation—not only of quality, but also of the inspection activity itself. A systemically controlled improvement process can be implemented via big data systems (essentially digital KAIZEN). This opens up further optimization potential. The systems are capable of analyzing the enormous volumes of data and thus recognizing complex interrelationships. In the future, robots will also become capable of learning through AI (humanoid robots). This is also due to new intrinsic capabilities. As a result, they will be able to act largely autonomously in the future. Initial examples already exist, such as at BMW.

But that's not all. Sensitive robots are becoming ever more "sensitive"—meaning they can take over more and more tasks from humans and interact with them, see Fig. 7.2. This development is further enhanced by the mobility of the new generation of robots. Through the digital twin (see next section), robots are also provided with the knowledge of what they need to assemble and how to do it. This also applies to autonomous transport units. They too "know" which part is needed for which product and when. The results are stored in the digital product memory and are available both for local and global production optimization, and of course for identifying patterns in current or upcoming quality issues. Digitalized logistics (see Sect. 7.1.4) provides notification if a part is delayed. Any necessary rescheduling can be carried out either via the global planning tool or by the individual networked intelligent machines.

The examples in Fig. 7.2 show that this is not a utopian vision. Here, for example, robots are used to install a dual-clutch transmission at Daimler. This application example was introduced several years ago. Almost classic areas of application are overhead work. When installing a door, the robots demonstrate their sensitive side. Thus, we are only at the beginning of this development.

7.1.3 Digital Twin

PLM systems are gaining a central and thus significant importance in the smart factory. The digital twin—that is, the digital twins—should have their origin precisely there; see also Chap. 4. This is not "just" a 3D model. All properties and functions of the planned product (material, software, sensors, dynamics, numerical simulation data, compliance information such as hazardous substances, required tools and aids, all the way to inspection characteristics, NC programs, and bills of materials) must be stored, see Fig. 7.3 and Chap. 5.

If the properties, such as functionality or geometry of the part or product, change, the effects on production and logistics can be determined immediately.

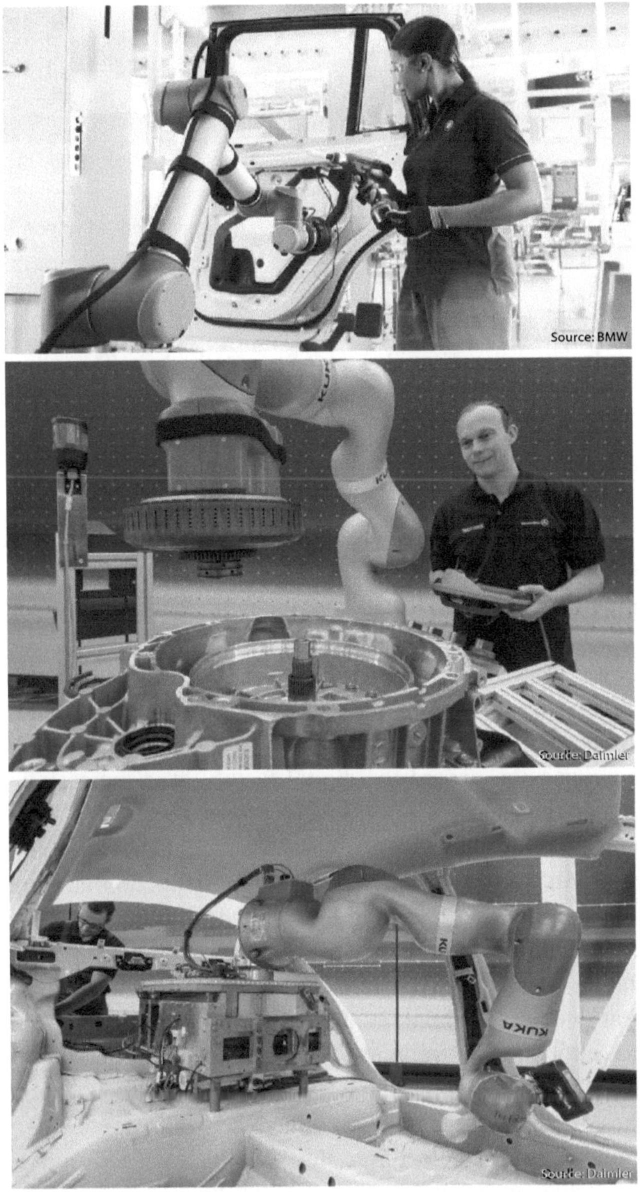

Fig. 7.2 Use of sensitive robots in automotive production

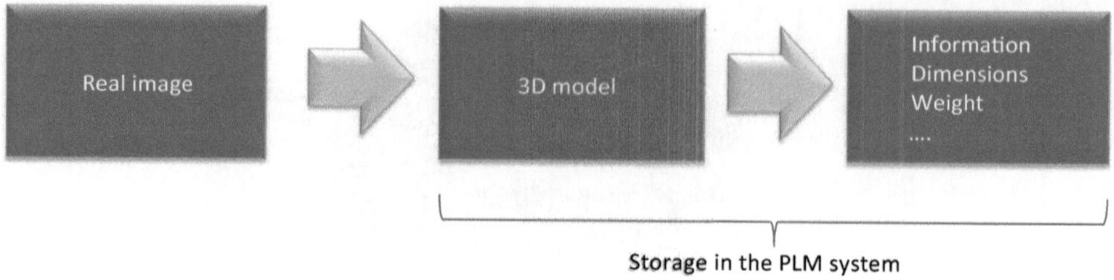

Fig. 7.3 Digital twin

Given the abundance of data, it makes sense to subdivide it. Thus, the following terms are used:

- Digital Product Twin,
- Digital Production Twin (production process, infrastructure model/machine model, automation model—everything related to SCADA, material flow simulation, inspection plans, production aids, NC program, inspection plans),
- Digital Performance Twin (e.g., throughput time, production time, quality).

This differentiation was already introduced in Chap. 5. Now it is being placed in the specific context. The first two points in the list are largely static in nature and should be stored in the PLM system. In the past, the PLM system was the domain of product development. With the expansion toward the digital factory, it is increasingly becoming the playground of engineering.

The last point in the list represents the digital product memory. As already mentioned, this is stored in the MES or big data system. Here, a distinction must be made between a digital twin and a digital shadow. In a digital shadow, communication occurs in one direction, whereas in a digital twin, it occurs in both directions. As a result, the complexity of implementation—and also its benefits—increases in the latter case; see [3].

It becomes clear that implementing all these described points is no simple undertaking. Therefore, it is not surprising that various product manufacturers (including Siemens) have developed corresponding development models; see Fig. 7.4.

The approaches to the digital twin presented here are not merely visions of the future. German automotive manufacturers, for example, are actively working on these approaches. 3D modeling and digital factory concepts have been established in this industry for quite some time; see Chap. 5. For example, the Austrian supplier Magna Steyr introduced a system for digital validation from process planning to production start as early as 2012. Audi built its latest plant in Mexico first in the digital world and then in the real world. In 2017, Gartner estimated that by 2021, around 50% of all companies would have implemented a digital twin. Such forecasts depend heavily on the definition of a digital twin. In recent years, this picture has changed significantly. Thus, it is not surprising that this forecast is no longer fully accurate and must be adjusted to reflect the

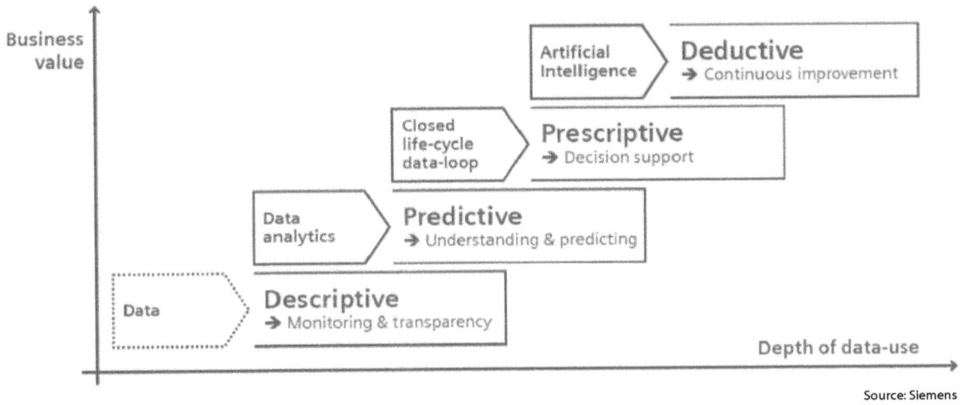

Fig. 7.4 Maturity model

insights and experiences gained. Regardless, productivity increases of at least 10% can be assumed; see [3].

The examples show, however, that integrating the individual systems can achieve a significant increase in process reliability and unlock corresponding economic potential. Furthermore, the real factory can be controlled and optimized via its digital representation.

7.1.4 Full Transparency through Logistics 4.0

In logistics as well, hardly "one stone is left upon another." The need for optimization is not only driven by cost considerations, but also results from the increasing individualization of products and the associated rise in complexity within logistics. Another factor is the heightened demands on delivery times, which must be continually reduced. This creates a conflict of objectives with the requirement to keep inventories as low as possible. However, current developments (chip crisis, COVID-19, political crises, cyberattacks) are also leading to a countertrend toward safety stocks.

The previously mentioned RFID tags provide significantly increased transparency throughout the entire logistics chain. Amazon, for example, is testing autonomous transport in the form of drones on the "last mile" to the end customer. The automotive industry is also active in this area. Audi is testing the use of these winged helpers in practical operations, see [3], [4] and also Fig. 7.5. Less futuristic is the use of AGVs for automated material supply and automated warehouses. The latter have now become part of everyday business operations.

The goal is to become ever faster, more flexible, and more efficient. This is also due to the fact that logistics costs and efforts, in the Lean sense, represent waste—necessary but non-value-adding activities that must therefore be minimized. Robots can take over

Fig. 7.5 Spare parts supply by drone

a large portion of warehouse activities (see notes on automated warehouses), and where human interaction is required, smart glasses support employees. Such systems are often met with high acceptance among staff. The key is to overcome the initial adjustment period.

Overall, consistent digitalization can reduce transport and storage costs by 15–30%. For inventories, leverage of 35–70% is even possible. At the online retailer Zalando, routing algorithms for order picking in the warehouse have also been optimized, resulting in a 10–15% improvement, see [1]. These examples illustrate the potential for optimization.

Of course, autonomous trucks are a major topic in the industry. Especially for small transporters, the driver accounts for about 30% of total costs, see [1]. Utilization of one's own fleet can be significantly improved via platforms. Average utilization is currently around 60%. Cloud-based approaches have therefore been used in the industry for years. In addition to optimizing the fleet, they are used, among other things, for tracking cargo. The introduction of autonomous trucks is delayed, making forecasts in this area extremely difficult.

All these activities and technical possibilities result in a complete integration of logistics and production, both for production planning and for operational execution. The outcome is holistic planning and thus further improvements.

7.1.5 Digital Supply Chain

The repeatedly mentioned transformation in value creation naturally also affects the entire supply chain. Its content is changing and it is being digitalized, see Fig. 7.6. Only the core processes are shown here. Supporting areas and processes are omitted for simplicity.

The topic of supply chain involves the planning, control, execution, and monitoring of both cross-company and internal value creation processes, including logistics and production—and thus the smart factory. The focus is on seamless material and value flows, and increasingly on information flows, from the identification of demand to its delivery to the customer. All participants in the process chain, from supplier to customer, must therefore be considered, including one's own company and its various areas and departments. Strictly speaking, the classic supply chain is transforming into a network of suppliers; the same applies to value creation—it too is evolving into a value creation network. In this context, the term smart supply chain is often used, see Fig. 7.6 and [7]. It is important that data silos do not hinder the changed value creation. These must be overcome at all costs. This point is of central importance, as data are the linchpin, both within one's own company and for the entire network. Technologically, this involves vertical and horizontal integration, see Chap. 5. This point has been mentioned several times—but on the other hand, it cannot be emphasized often enough. Without consistent and correct data, nothing works anymore.

A key element of Industry 4.0 and the associated digitalization is that positive effects can be achieved both in batch size 1 and in the area of economies of scale. Companies

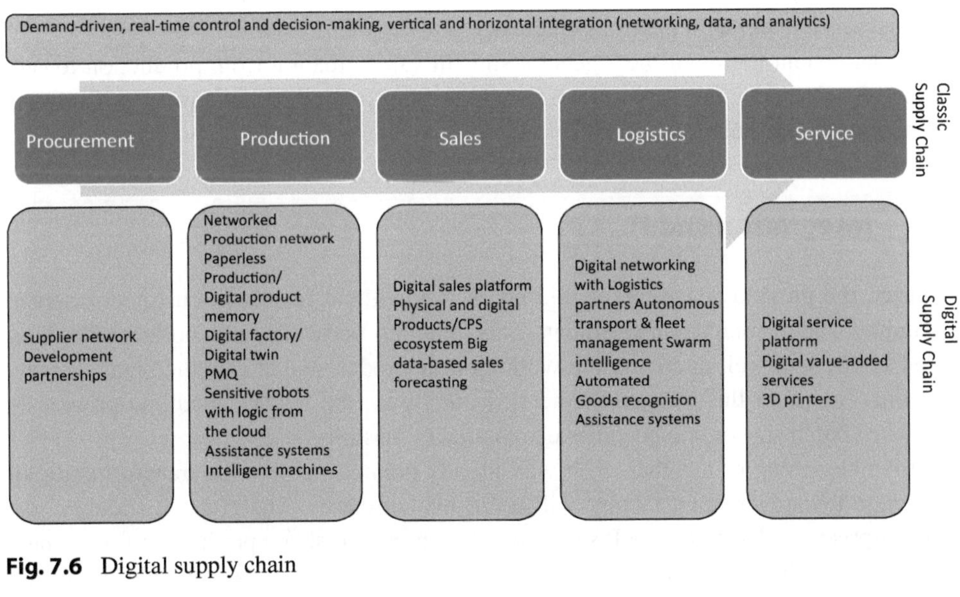

Fig. 7.6 Digital supply chain

can increase the efficiency of their existing value chains or enhance the flexibility of their own production systems, thereby unlocking entirely new business models. Value creation will take place among actors who are much more interconnected than today. This also means a greater dynamization of the value chain (for example, to respond to changing customer requirements). Furthermore, resilience must be increased. The recent crises (COVID-19, political influences, chip crisis) show that such an environment is becoming the norm. The fundamental value creation opportunities will be significantly expanded by digitalization and Industry 4.0. As a prerequisite, companies must be aware of which areas of their value creation they want to change, in what way, and what is possible under market conditions. Admittedly, the spectrum of change is broad and extends to entirely new business models.

Another very significant driver of change for the entire value creation process is the 3D printer (i.e., additive manufacturing). This leads to changed requirements for suppliers and their parts. Whereas previously suppliers were required to deliver prefabricated parts in the right quality, on time, and at the right price, these requirements will at least partially shift toward pure raw material supply.

The starting point for additive manufacturing is powder made from the target material (such as metals). The associated logistics will not change in the next few years, but in the long term, significant changes are expected. Instead of transporting parts, CAD and 3D models will likely be exchanged in the future—without trucks. This could even extend to self-service, where the customer prints the required parts themselves on their home 3D printer. The customer acquires the necessary data for "printing" instead of purchasing the part. This is already possible today and is already being practiced. In practice, however, the main obstacle is often simply the lack of the required 3D models. In general, additive manufacturing processes will complement existing approaches. This topic has already been discussed in detail in previous chapters.

To what extent there will be a general shift in value creation from production toward product development remains to be seen. However, product development will clearly become more important for companies.

7.2 Integration and IT

However, the previous measures would be useless without real-time, customer-oriented scheduling and resource management. This requires vertically and horizontally integrated systems, as well as changes in working time models and streamlined management structures (more on this in later chapters). In this way, the much-touted and previously mentioned batch size of one becomes economically feasible.

At the beginning of the chapter, it was already pointed out that the transformation of production towards a smart factory will occur evolutionarily. Therefore, it should come as no surprise that MES and ERP systems will continue to shape production for the next several years. The novel CPPS are currently not yet commercially available or are not mature enough for practical use. As a result, master data management will remain the responsibility of the ERP system for the foreseeable future. In many companies, this master data management is still not at the desired level of quality and sophistication. Without "good" data, there can be no efficient digital processes. The rule is: the more digital a production environment, the more important the data becomes (as has now been repeated almost like a mantra). This point must not be underestimated. Paper-based processes differ significantly from their digital counterparts, particularly in terms of data usage. While data quality issues in paper-based processes can be remedied through manual intervention, having the "right" data is essential in digital processes. Achieving high data quality requires a cultural shift within the company, which takes time and consumes a great deal of energy.

To support transparency on the shop floor, the already frequently mentioned RFID-based approaches should be implemented. RFID is developing into an "Industry 4.0 base technology." RFID tags are often seen as a place to store any kind of information (such as a digital product memory). However, their storage capacity is currently still quite limited. They can become "smart" by linking them with an MES or the cloud. Thus, MES will retain a certain justification for existence for the next several years. In addition, it is essential to ensure seamless connectivity of production equipment/machines and other equipment. This can help avoid manual activities (such as documenting machine downtimes).

The integration of PLM systems with an MES should not be underestimated. The added value, for example in the form of the digital twin, can only be realized with seamless integration of both systems. Otherwise, the whole thing makes little sense.

7.3 The New Role of Humans in Production

Due to the increasingly supervisory role of humans, people are becoming the "conductors of production." Of course, there will still be people in the smart factory performing classic production tasks. Maintenance, too, will not be possible without people carrying out repairs on equipment, regardless of all optimization approaches. It is especially in the area of maintenance that the degree of change—and thus the need for qualification—is most apparent. Not only will sensitive robots support employees in production, but assistance systems in various forms will further optimize work, and novel tools such as exoskeletons (for example, the Chairless Seat at Audi) will ensure that employees can remain productive even at the age of 60. Virtual reality approaches can also be used for employee training and maintenance training. This results in much steeper learning curves and allows for the coverage of rare special cases in practice, see also Chap. 5. In addition, mobile solutions, such as for material withdrawal for maintenance measures, help to both relieve employees and simultaneously improve data quality.

Regardless of the degree of automation, the combination of humans and sensitive robots offers the potential for significant labor cost reduction. Details have already been described in Chap. 4, see [6]. Furthermore, transparency increases, as robots provide a wealth of information about the production process. An interesting example of the implementation of a smart factory is certainly the aforementioned Factory56 by Daimler.

It is not enough to simply introduce new technologies; all participants must also master them optimally. Therefore, continuous further training is indispensable. To achieve this, the organizational structure of the company must change, and career models must also adapt to the digital transformation. These points are addressed in Chap. 9.

The examples and scenarios presented are intended to provide an impression of current and future developments. The shift toward digital production cannot be stopped, if only for cost reasons.

References

1. Jürgen Meffert, Heribert Meffert, Eins oder null, Econ Verlag, April 2017.
2. Britta Mittlefehldt, Harald Proff, The Smart Factory: Industrie 4.0 unter realen Bedingungen erproben, Deloitte, 2023.
3. Christy Pettey, Prepare fort he Impat of Digital Twins, Gartner, 18. 09.2017, https://www.gartner.com/smarterwithgartner/prepare-for-the-impact-of-digital-twins. Aufgerufen am 18.11.2017.
4. Dietmar Poll, Drohen in der Intralogistik: Geflügelter Lieferservice, Produktion, 30. März 2017.
5. Philipp Vetter, Audi schafft das Fließband ab, Welt 24, 21.11.2016.
6. Horst Wildemann, Produktivität durch Industrie 4.0, 24. Münchner Management Kolloquium, 14./15. März 2017.
7. Mario Zillmann, Keine Industrie 4.0 ohne Digitalisierung der Supply Chain, Lünendonk, 2016.
8. https://www.industrie40award.com/ Aufgerufen am 04.06.2023.
9. https://group.mercedes-benz.com/innovation/produktion/factory-56.html. Aufgerufen am 04.06.2023.

8 Opportunities Through New Business Models

The digitalization of products and production alone is not sufficient to remain competitive in the future. These serve to increase efficiency, reduce costs, and enhance flexibility, and are thus primarily defensive measures. It is therefore necessary to advance the shift from defensive to offensive action. This step leads to new digital business models, see [16], [27].

When it comes to business models, the focus is on how and in what way companies earn their money. This essentially includes who pays for a product or service and in what form. This view, which is quite straightforward for traditional products and services, becomes significantly more complex with digital business models. Especially in the B2C environment, users are often provided with services seemingly free of charge, such as online editions of magazines and newspapers. The provider receives compensation indirectly, for example through advertising revenue from third parties or the use of user data, see [1]. To put it clearly—as a customer, you pay with your data. New business models thus emerge because smart products, with their expanded or altered range of functions, enable novel usage offerings and thus new business models.

There are two approaches to changing one's own business model:

- evolutionary change and thus further development of existing business models,
- disruptive change and thus completely new business models.

Companies can certainly pursue both approaches in parallel—as a hybrid approach. In general, the hybrid approach is even the more sensible one. Amid all the enthusiasm for new possibilities, the "classic" sale of products, especially in mechanical and plant engineering, should not be neglected. Such models will still have their justification in the coming years. In the next few years, traditional business models and their optimization in the form of evolutionary changes will certainly account for the majority of total revenue

W. Huber, *Industry 4.0 Compact—How Technologies Are Transforming Our Economy and Our Companies*, https://doi.org/10.1007/978-3-658-51029-9_8

for traditional companies (of course, there are exceptions here as well). However, new approaches will gain relevance over time, see [14]. Thus, companies are well advised to initiate corresponding activities in a timely manner.

In the often-mentioned disruptive business model, the nature of the service or product changes completely, meaning new providers enter the market and/or existing products or services are completely replaced by new ones. In Chap. 4 the example of photography was already mentioned. Incidentally, Kodak, once a global corporation, had to file for bankruptcy in 2012. Currently, the company is successfully active in the printer business. During the same period, Instagram, a platform for editing digital images, was sold to Facebook for around US$1 billion. At the time of the sale, Instagram had about 20 employees, see [1]. This is not an isolated case of an established company failing. A company history of about one hundred years is no guarantee of successful continued existence in the future.

Unfortunately, there is no uniform and widely accepted definition for a disruptive business model. Disruptive changes also occur much less frequently than evolutionary ones. Amazon, for example, has not developed a disruptive business model. Customers also pay with money at Amazon, and the goods sold are not significantly cheaper. However, the company has perfected online retail. Of course, one can argue whether the digitalization of commerce constitutes a disruption or not.

The interesting thing about this example is that successful companies often cling to existing (and often still profitable) business models for too long. They fail to recognize the "signs of the times" and are mercilessly overtaken by time—that is, by new digital and disruptively operating competitors. This pattern is called the "innovation dilemma," see [1]. As already mentioned, Kodak is not an exception here. The hotel industry has been experiencing a similar digitalization with Airbnb for several years. Quelle, as a classic mail-order company, was successfully challenged by Amazon. The telephone industry and music industry are further examples. The automotive industry is also facing new competition from challengers like Tesla. However, the example of the automotive industry also shows that established companies can continue to be successful in the market. The examples mentioned demonstrate that digitalization has affected all industries. Only the degree of change varies from industry to industry, which certainly also has a temporal component.

The economic relevance of digitalization was already addressed in the first chapters. There are numerous studies on the economic significance of the topic and a multitude of practical examples. Regardless of the sometimes widely divergent statements and forecasts, the economic relevance is undisputed. A study by the Future Council of the Bavarian Economy speaks of 40% more job growth and 80% more revenue with the use of digitalization over the next few years, see [23]. According to a study by the consulting firm Accenture on behalf of the Austrian Ministry for Digitalization and Economic Location, digitalization can increase gross domestic product by up to 1.9%, see [12]. It is not just about making one's own company fit for the future in terms of flexibility. Some authors also refer to "robust companies," see [29, 33]. This robustness is becoming increasingly important, as crises are becoming more frequent. In this context, it is also important to question one's own business model and to what extent digitalization and Industry 4.0 offer new opportunities for innovative business ideas for one's own company. Other sources predict annual revenue growth of 7–10% for companies through the development of new business models. Regardless of the respective studies and investigations and the associated figures and forecasts, the shift towards Industry 4.0 and digitalization represents a key to maintaining competitiveness in Germany as a high-wage location, see [21, 30, 34].

In general, efforts should be made to invest in new business models according to one's own capabilities. Optimizing existing business models is certainly to be welcomed as a first step. Thus, industrial services that can be implemented quickly and easily should be realized at the beginning. In this way, fundamental skills and experience can also be built up gradually. At the same time, initial successes are quickly visible and change management is also manageable. Thus, the first service offerings represent a supplement to the existing products. Market feedback and, if necessary, modifications to the range of services can also be implemented quickly. Investments and the associated risks are thus also kept within limits.

GE has established new business models based on its cloud platform Predix. The company invested around US$1 billion for this purpose. The platform is used, for example, to optimize wind turbines for the power plant operator E.ON. Using sensors, the systems can be maintained accordingly and their operation optimized. According to GE, US$800 million in revenue had already been generated through Predix by 2014. However, in 2019 the company sold the platform because revenue expectations were not met, see [35]. The logistics service provider UPS has optimized its business model using sensors. They are used to locate and optimize delivery vehicles. According to McKinsey, such approaches can save around 17% of logistics costs by 2025, see [15].

Zeppelin also aims to generate around 50% of its total revenue from digital transactions by 2025, see [7]. Thus, it should come as no surprise that, across industries, B2B business volume is significantly higher than that from the B2C segment, see [14].

Viewed objectively, the customer does not care how a product is created. Therefore, the value of the product or service for the customer should be the focus. The value for a customer is extremely subjective. It consists of the benefit and the image. This applies to both B2B and B2C business. That is why companies like Uber and Airbnb are so successful. They deliver value to customers; they benefit the customer, see [1]. However, in recent years, both companies have also experienced their lows. This shows that digitalization is not a panacea. Digitalization is therefore a means, not an end in itself.

Regardless of all discussions, unfortunately there is no universal recipe for developing digital business models. A value-centered approach is certainly the most promising. This means consistently aligning one's own business with customer needs and the customer experience. This "wisdom" does not originate from the digital world, but has long been an important guiding principle in the Lean world. Ultimately, a coherent picture must be created for the customer. Developing a new website or unmotivated apps only incurs costs. Furthermore, the existing criteria for sustainable customer satisfaction must not be neglected:

- Quality,
- Cost,
- Delivery date and delivery reliability,

must not be overlooked. They also apply in the digital world.

In online retail, very basic rules apply. Loading times are a key factor in customer satisfaction. If delays occur here, even in the range of seconds, they have a significant impact on sales. If there are delays of 10%, this results in a sales decline of approximately 2.4% per visitor and an overall sales drop of around 4.2%. It gets even worse. Customers who experienced longer wait times were less likely to return to the website on the same day. It is also important that retailers make the payment process as simple and convenient as possible for customers, see [7]. Amazon is certainly a role model here,

8.1 Backgrounds for Change

For a long time, German companies—and especially the German Mittelstand—were protected from competitors (mainly from low-wage countries) by their technical knowledge advantage regarding their products. Due to increasing digitalization and the growing prevalence of platforms, this position is now increasingly exposed to a multidimensional threat. The challenge now is to transform the domain knowledge built up over years or even decades in the analog world into the digital world.

Digitalization is fundamentally transforming the entire value chain and established business models, see Chap. 7. Existing products and offerings are increasingly being supplemented or even replaced by digital and online services. For example, if companies know at any time whether and how customers are using products, they can develop entirely new business models, such as pay-per-use, e.g., in the engine sector at Rolls-Royce. Furthermore, manufacturers can stay in contact with customers through the products. This forms the basis for a direct and continuous dialogue, including direct observation of customer satisfaction.

The change in value creation can be well illustrated by the example of print media. For a long time, this sector was able to generate a very profitable business model through the sale of newspapers and magazines, including advertising revenue and income from classified ads (job ads, real estate, cars, matchmaking, etc.). With new market entrants such as the Scout Group, Parship/eDarling (and other matchmaking services), eBay, etc., the classified ad market collapsed rapidly. In addition to reduced sales due to increased online offerings and the significantly decreased advertising revenue, this led to the demise of some renowned daily newspapers, such as the Financial Times Deutschland. The industry has changed significantly in recent years. Digital offerings have supplemented print offerings. Additional revenue from advertising could be generated. This dramatic upheaval is still ongoing. AI will cause the next change, as articles will no longer be written by editors. However, AI cannot replace research. The newspaper industry, as an important source of information for all of us, demonstrates the dramatic change in the underlying business models.

The business model is also changing because there is a stronger focus on service and thus away from product ownership. Thus, differentiation is increasingly achieved through the offered service and less through the physical product. As has been mentioned several times, these services are provided in digital form. In the context of one's own ecosystem, these services can be provided by the company itself or by partners.

This is also evident in the potential that lies in sales and service business:

- Generation of an information advantage,
- Cost reduction,
- Reduction of process complexity,
- Improvement of effectiveness,
- Increase of brand awareness and improvement of image,

see [5]. Thus, it is not surprising that in the future there will probably be no B2B sales without online interaction, and thus B2B sales will become much more similar to B2C sales. The customer interface will therefore continue to play a central role in future value creation networks. In contrast to today, the interface will increasingly be digital. Controlling this interface therefore has decisive strategic importance for companies, see [5]. Whoever controls the customer interface controls the market (digital point of sales).

In contrast to production and logistics, digitalization approaches are much easier to implement in sales and service. However, the benefits are much harder to verify. The importance of digital marketing platforms is still often underestimated. Amazon and others have vividly demonstrated their significance. This also applies to B2B sales. In the European and German context, Gaia-X, Catena-X, and Manufacturing-X are trying to establish themselves, as are commercial providers such as Otto or Zalando. The two commercial examples show that such attempts can indeed be successful, even if not comparable to Amazon. Overall, the new possibilities of digitalization must be leveraged not only to achieve cost advantages, but above all to generate competitive advantages. Customer experience management (multi-channel approach) is particularly noteworthy here. This provides the opportunity to better address customer needs as well as to assess the customer value of one's own solutions.

Regardless of the new possibilities, corresponding customer benefits must be created (regardless of whether the customer is an end consumer—in the B2C environment—or a company—in the B2B environment). In the end, the customer must buy and accept the new or changed products, including the business models. There must therefore be a value proposition. In the B2C environment, these are, for example:

- making life easier and more comfortable (for example, through smart lighting or radiators),
- detecting problems in a timely manner, as with smart devices that detect health issues or the previously mentioned sensors in machines,
- generating economic benefits, for example through a changed cash flow.

Currently, the rapid increase in technical possibilities (such as the continuous increase in computing power) is creating new opportunities. This, combined with the steady spread of individual technologies such as the smartphone, is consequently leading to changed consumer behavior. However, the example of GE's Predix also shows the downsides. Misinvestments, such as insufficient focus, can lead to the failure of the project.

8.2 Characteristics of Digital Business Models

The fundamental logic of business models (i.e., the offering of services) is, of course, not nullified by their digitization, regardless of whether an evolutionary or a revolutionary/disruptive approach is taken. Unfortunately, there is no clear and precise definition or accepted understanding of the term "digital business model." A pragmatic definition is that the transition to a digital business model is complete when digitization permeates the entire value chain and is not merely focused on marketing or a more or less "attractive" website. Companies have thus implemented the transformation toward a digital supply chain (see Chap. 7) and consequently use both revolutionary and evolutionary technologies to map all relevant transactions through it.

According to Gartner, there are six stages in the development toward a digital business model, see Fig. 8.1 and see [22, 36]. This model ranges from analog approaches to fully autonomous business relationships. The speed of development of an evolutionary versus a disruptive approach is evident, as is the destructive impact of the disruptive approach. In this context, several stages of development can simply be skipped. However, it has also become apparent in recent years that disruptive approaches account for a significantly smaller share than was predicted a few years ago, see [11].

In practical implementation, digital business models differ significantly from "analog" ones (as also shown in Fig. 8.1). For example, selling music CDs in a store is a different model than selling via the Apple Music Store iTunes. The two approaches also differ significantly for the consumer, since in the above example, the entire CD no longer needs to be purchased; instead, only the personal favorite track can be bought, or the purchase can be skipped entirely in favor of acquiring a time-limited right of use. The situation is different from the consumer's perspective in video rental. Here, the service offering differs less. In both the analog and digital worlds, a video/DVD is rented.

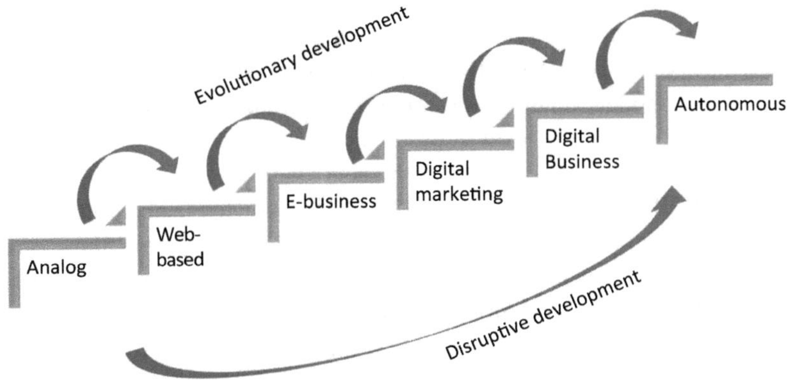

Fig. 8.1 Gartner model for the 6-stage development of business models

The way of renting, of course, differs. But even here, films can be acquired for one-time viewing.

The distinctiveness of digital business models compared to analog ones is due, among other things, to the different technologies used (see Chap. 4 and 5). In addition, all activities are digitized, from service delivery to payment. The basis can, of course, still be machines and equipment.

An important factor is the unlimited scalability and expansion of digital business models—i.e., virtually zero marginal costs. Through digitization, products are partially or completely dematerialized (e.g., eBooks versus printed books).

It should be noted that digital business models in particular are not permanent. They change due to the market, customers, rapidly evolving technologies, and the ease with which business models can be copied. As a result, they change much more quickly than analog ones, as seen in the music industry or print media, as previously described.

It can be observed that market requirements and customer needs diverge in macro and micro markets. This leads to increasing complexity in the supply chain, up to the atomization of the value chain. In addition to the ever-increasing fragmentation of the value chain, the speed of product innovation is also rising, see [31]. In this context, the term digital Darwinism is also introduced, meaning that technologies and requirements change much faster than companies are able to adapt. Thus, it is not the largest or most profitable company that will survive, but the one that is best and fastest at adapting to rapidly changing conditions. It should be noted that "speed" and "time" are highly industry-dependent and therefore relative. The aerospace industry, for example, has very different innovation cycles than the chip industry. This is due, among other things, to the complexity of the products and regulatory requirements.

It has already been pointed out several times that digital transformation blurs industry boundaries, but also those within one's own company. As a result, partner management takes on strategic importance and must therefore be an integral part of corporate strategy. The company's own monitoring radar must therefore be expanded. Competitors no longer arise only from the circle of "known" challengers. More on this in Chap. 10.

Generally speaking, the classic B2C business is much more affected by the digitization of business models than, for example, the B2B business. Here, optimization of internal processes, often in the form of production and logistics, is usually the first step, driven by digitization and Industry 4.0 approaches. However, it is also worthwhile to view one's own B2B business through a "B2C lens." Knowledge about the end customer (more precisely, their use of one's own products) helps to improve these sustainably. Thus, the boundaries between B2B and B2C business are increasingly disappearing.

Regardless of the specific form, the focus should be on one's own strengths and thus on core competencies. If these are transformed into the digital world, the lead over attackers and new market entrants, such as start-ups, is maintained.

8.3 Impacts on Existing Business Models

Digitization affects all elements of the value chain. These effects will be very significant. In Chap. 7, the digital supply chain was already introduced. It provides an impression of the upcoming changes. Every industry will be affected by these changes. Entire industries can be disrupted or even rendered obsolete. One industry currently experiencing very significant upheaval is logistics. It is undergoing massive changes from various directions, see Fig. 8.2.

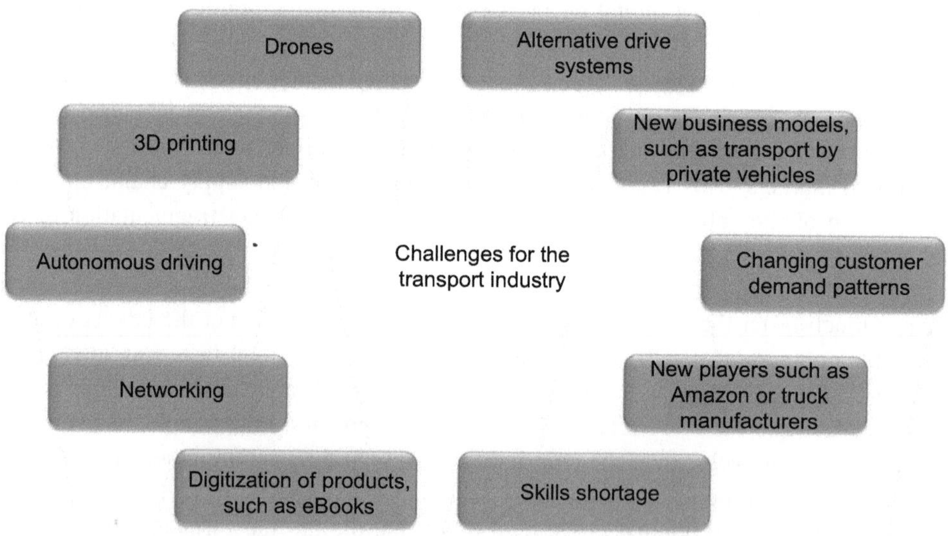

Fig. 8.2 New challenges for the transport industry

For example, Amazon has a license to charter and load ships. In addition, in 2016 the company leased around 40 Boeing 767s to ensure sufficient freight capacity, see [18]. At the end of the day, logistics service providers risk becoming obsolete.

Therefore, it is necessary to continue optimizing one's own business model and to identify new business areas. In general, the following potentials for optimization arise:

- Increase in efficiency,
- Improvement of the decision-making basis.

see, among others, [15]. The points listed, of course, apply regardless of whether existing business models are further developed in an evolutionary way or new disruptive/revolutionary approaches are chosen. Both points affect smart products as well as the operational execution of business processes.

8.4 Evolutionary: Further Development of Existing Business Models

In evolutionarily developed business models, the existing and established rules of the game in their respective industries remain valid. New technologies primarily serve to further optimize the model. Thus, only gradual changes are made to the business model elements. The company's core competencies and product portfolio are also not changed, but are merely expanded and optimized accordingly.

Characteristics of the evolutionary approach are:

- Focus on technological and process efficiency improvements,
- digital optimization and thus enhancement of product functionalities and services,
- product-related digital services,

see [6]. According to McKinsey, this approach will account for the largest share of total revenue among established companies in the coming years, in contrast to disruptive approaches, see [14]. The susceptibility to change also depends on the time cycles, which, as already mentioned, vary from industry to industry. Thus, for sectors such as mechanical and plant engineering as well as the aviation industry, the evolutionary development of their own business models is likely to remain the focus.

Currently, marketplaces are experiencing a digital renaissance. These are operators of digital platforms that enable the trading of services and products. As a result, retail is changing. Multi-channel approaches allow for further increases in revenue. Here, consumers are offered tailored products both via smartphones and in-store. Walmart is investing heavily in digitalization and, in recent years, has been the only established American retail company to achieve double-digit growth rates. Competitors such as Sears, on the other hand, have suffered significant losses in market value over the same period. Walmart sells around 67 million products online and ranks second in this market segment in the USA, ahead of eBay. Conversely, Amazon acquired the organic supermarket chain Whole Foods for $14 billion in order to build up its own brick-and-mortar business. At present, a battle is raging between the established company Walmart and its challenger Amazon. The gap between the two competitors in terms of revenue has narrowed significantly, from $363 billion to $116 billion in 2016 and $329 billion to $51 billion in 2012, see [19], [38]. In 2020, Walmart achieved revenues of $524 billion, making it number one on the Fortune Global 500 list of the world's largest companies by revenue. The company has seen disproportionately high growth, especially in online business, see [37]. The race between these two giants thus remains exciting.

8.5 Disruptive: Development of New Digital Business Models

As already mentioned several times, the disruptive approach involves a radical change, that is:

- radical change in value and service orientation,
- use of new disruptive technologies,

see [6]. The shift from product providers to service providers has been discussed repeatedly. Printer manufacturers are just one example. Currently, hardly any large printers are sold; instead, their use is billed via corresponding fees. This is made possible by sensors that record actual consumption or usage. The same applies to the company Kaeser (a manufacturer of compressed air compressors). The company has switched its business model to selling compressed air. Here too, sensors are used, see [15].

Although experience with digital business models has not yet become widespread, some cross-industry patterns are emerging:

- strong growth,
- network effects up to and including "the winner takes it all",
- zero marginal costs,
- global availability, rollout in third countries can be carried out very easily,
- immediate availability, for example by downloading the product,

see e.g. [16]. These patterns do not necessarily apply to all such approaches.

New business models are highly dependent on the respective industry and the products offered. However, some drivers (economic and technical) and characteristics for new digital business models can be identified:

- personalization and individualization,
- CPS,

- service instead of product,
- success-based approaches,
- (additional) sale of collected data and use of external data,
- cloud/platform
- digital ecosystems.

Naturally, the individual characteristics should not be viewed in isolation. In practice, there is usually a strong blending of these elements. This is also evident in the examples provided.

Regardless of the specific driver, it is important for companies to be the first to establish new business models. This is of fundamental importance for securing the future, see [28, 30]. This requires a constant questioning of one's own actions and thus a creative "destruction." The guiding principle is: "If we don't create the thing that kills us—someone else will do it."

However, previous experience also shows that, especially in the B2C environment, complete digitalization does not represent the optimum, as the example of Amazon demonstrates. Experience in recent years has also shown that, in many cases, evolutionary developments are the dominant element. By nature, disruptive changes appear much less frequently on the economic stage. But when they do occur, they unleash their full power.

8.5.1 Personalization and Individualization

The automotive industry vividly demonstrates individualization. Above all, German manufacturers offer their customers an almost unlimited variety of options. The same trend can now be seen in sports shoes. Here, the soles are adapted to the specific fit of the wearer using a 3D printer. These examples can be extended almost indefinitely. Thus, every company is encouraged to validate individualization strategies and, if necessary, implement them as quickly as possible. This applies especially in the B2C environment, but increasingly also in B2B business.

In retail, customers are offered individual deals tailored to the product they have just purchased, such as a new pair of jeans or a dress. Amazon has been successfully pursuing this strategy for years. Especially in the B2C segment, it is evident that multi-channel approaches are clearly superior to purely digital approaches. In the USA, around 70% of brick-and-mortar business was digitally influenced in 2015. This trend has intensified since then. American retailers are much more creative here than their German competitors. The department store chain Macy's invested considerable money in its cross-channel activities. Retailers found that customers who shop at Macy's through multiple channels generate significantly more revenue than customers who are served exclusively through a single channel, see [18]. Using AI approaches, customers in the B2C environment are offered individualized deals.

8.5.2 CPS

Intelligent machines and systems have already been sufficiently discussed in Chap. 4. Through sensors and actuators, not only does the boundary between the real and digital worlds disappear, but they also serve as the starting point and often the prerequisite for new business models. With CPS, it is possible to save about 2–5% of total fuel consumption in aviation. The potential is significantly higher in the container sector. Through integrated tracking, between 10 and 25% of global containers could become redundant, see [14].

With detailed knowledge of the actual use of products, manufacturers are able to proactively offer suitable services to customers. For example, if a truck with a capacity of 20 t is consistently used with only a 5 t load, there may be optimization potential. The same approach can be applied to vehicle utilization. Analogous approaches exist in mechanical and plant engineering. Here, unplanned machine downtimes are the main lever for optimization and thus for new business models. This approach also applies to wind farms, oil, and gas production platforms. They are monitored by sensors, and failures can be proactively prevented. In the event of a problem, remote maintenance enables a thorough analysis. This allows costs to be reduced, productivity to be increased, and plant availability to be improved, with a total potential of around 160–930 billion USD in 2025, see [14].

These condition-based approaches (specifically in the form of Condition Based Maintenance and Quality) create new opportunities to expand service offerings in virtually all areas. The prerequisite for this is the combination of sensors/actuators with software and their communication capabilities—that is, CPS. The possibilities described above overlap significantly in their specific manifestations—but this is inherent to the subject.

8.5.3 Service Versus Product

Services can either replace existing products or offer additional features through them. Making phone calls via WhatsApp or Apple Facetime is gradually replacing the traditional telephone and serves as an example of new services. In office environments, telephones are increasingly being phased out. Here, services such as Microsoft Teams are being used more frequently. In all these examples, only an internet connection is required. Marginal costs are non-existent or extremely low. At the same time, new products can be created by simply combining existing functionalities.

The topic of data was already described in detail in Chap. 4. Data enables the generation of a multitude of new business models, see Sect. 8.5.5. In addition, digital services are immediately available anywhere in the world, see [17].

Another approach is the modularization of services. This allows the advantages of standardization to be combined with those of customization. These services can then be combined or expanded into customer-specific offerings, see [30].

As with the description of individual technologies, the new possibilities listed above are only partially truly new. When it comes to services instead of, or in addition to, products, the customer faces a significantly lower financial requirement, as only ongoing operating costs are paid. For the supplier, this leads to a continuous cash flow, including an additional source of income (from the service itself). As a result, responsibility for maintenance and proper operation of the systems often shifts to the manufacturer, along with the provision of the service.

A prime example here is owning a car versus using car sharing as a competitor or alternative. The car itself loses some of its significance. For instance, the number of registered users of car sharing services in Germany rose from around 450,000 in 2013 to over 1.7 million in 2017, see [2]. In 2020, more than 2.29 million users were registered with over 220 providers, see [39]. With data, networking is also becoming increasingly important. The commercial vehicle sector is certainly leading in this area. Many trucks today are connected to their environment and provide real-time data for optimization, such as remote and preventive diagnostics. In the passenger car sector, there is currently little willingness to use such services. However, car insurance companies are increasingly offering behavior-based insurance policies.

Such use of data can also increase customer loyalty, as manufacturers can better fulfill their value proposition. Furthermore, vehicle data provide OEMs with valuable and important insights for the further development of existing and new services or functionalities. These new services also give rise to new ecosystems. Larger operators and vehicle manufacturers will increasingly dominate the market. Small and independent workshops are therefore well advised to enter into appropriate partnerships, see [8].

Predictive maintenance and operation are a special case of additional services. According to McKinsey, this could save around 1.6 trillion USD worldwide by 2025, see [14]. With the service approach, the supplier is much more closely involved in the customer's internal processes. This provides opportunities for further (sales) activities and new business opportunities. In addition, continuous customer contact results in much stronger customer loyalty than, for example, a one-time product sale.

This approach is also pursued by the fittings manufacturer Grohe. The company spent two years developing a system for detecting damage in water pipes. Using sensors and a specially developed algorithm, the water damage warning system is able not only to detect current water damage, but also to identify impending pipe bursts and, if necessary, shut off the water supply, see [32].

8.5.4 Success-Based Approaches

A consistent extension of the above approach is represented by success-based models. For example, guaranteed availability of machines and systems provided by the service provider. In this case, the risks are transferred to the provider and service deliverer. This can significantly strengthen customer loyalty. Here, too, ongoing revenues are generated.

GE, among others, practices this approach. The company operates business-critical applications in this way. As a consequence, a high level of customer dependency is also created, see [15].

A success-based approach significantly reduces the investment risk for the customer. If such services are offered by established companies, they represent a very attractive alternative to existing approaches for customers, while simultaneously improving overall cash flow. Of course, providers charge for such models.

8.5.5 Selling and Using Data

The data generated by new products can also be valuable and of interest to other actors and companies. The "marketing" of weather data or providing robot data from one's own production to the manufacturer (see, for example, Kuka in Chap. 4) are just a few examples. Even residents' registration offices have discovered this as a new source of income. Catena-X and Manufacturing-X are also trying to build a data economy. Data provision can also be done anonymously, so that no conclusions can be drawn about, for example, the quality of one's own production. Combining one's own data with external data can create added value for both the company itself and other actors. The decisive factor here is the degree of data usage. The customer must give consent regardless of the degree of data usage. Customers can also actively share data with third parties. Especially in the B2C sector, this gives customers the opportunity, for example, to obtain more favorable insurance terms, as already mentioned.

Using external data goes in a very similar direction. In some cases, the approaches merge and blur. As an example, data from the telematics sector of road users can be mentioned. Other scenarios include linking elevator data with the locking system in hotels to optimize the required number of elevators. The source of data usage is therefore very diverse.

8.5.6 Platforms

The technology-centered perspective has already been discussed in detail in previous chapters. Platforms and digital ecosystems often go hand in hand. Nevertheless, an attempt is made to consider both topics separately—fully aware that there are overlaps.

Platforms represent the central technical foundation for the dissemination of smart services. They are offered both by large companies such as Amazon, Apple, BMW, Bosch, General Electric, IBM, Kuka, Microsoft, SAP, or Siemens, as well as by startups. The politically motivated platforms Gaia-X, Catena-X, and Manufacturing-X (as already mentioned several times) are further examples. The primary aim here is to limit the influence of American and Chinese companies. The list can be extended almost indefinitely. Despite sometimes very different characteristics, they all have one thing in common. Platforms effectively insert themselves between end customers and producers. In the first step, both sides are promised advantages. Through platforms, such companies build an ecosystem. For customers, the most interesting platforms are those with the largest and most diverse offerings (i.e., as many suppliers and services as possible). It is encouraging to note that the list of platform operators above also includes very renowned German companies. The politically motivated approaches, as already mentioned, are also to be viewed positively. However, these platforms do not offer only intermediary services, but also additional services. Examples include additional warranties, payment processing (in both directions as customer and supplier), free delivery, as well as big data solutions, etc. A distinction is made here between closed systems (such as Apple's, which functions exclusively via their operating system) and so-called open systems (such as Amazon).

Start-ups in particular try to win as many end customers as possible for their own idea in the early stages (the failure to achieve this goal is often the reason for the failure of such start-ups). In the B2C environment, this is often done through traditional advertising, such as TV commercials and on social media. However, especially in the B2C environment, a general problem also becomes apparent. (Especially long-term) customer loyalty can hardly be achieved through purely digital communication. The platform must compensate for this through image, functionality, and value proposition.

Operating a platform can therefore be a highly lucrative business. At the same time, it is essential to master the respective platform. For companies, it is therefore imperative either to build such a platform themselves (as the examples above show) or to do so with partners. Companies must do everything in their power to prevent a platform from inserting itself between them and their customers; otherwise, they risk being relegated to mere suppliers, exposed to corresponding price pressure and price dictates.

However, mastering a platform is not "set in stone." Suppliers can themselves form alliances and thus strengthen their position vis-à-vis the platform operator. Their own reputation is another lever for "boarding" an existing platform. However, this is no guarantee of success, as demonstrated by the music industry. New market entrants, such as Apple or Spotify, have overtaken established companies like Sony Music—and in a very short time.

From what has been said so far, it is not surprising that a "war of platforms" is looming, or rather, is already in full swing. Platforms will continue to "spring up like mushrooms." However, in the coming years, some platforms will lose significance, disappear, or there will be mergers of platforms.

Another driver for platforms is the redefinition of products. The automotive industry also provides a very good example of this. Companies such as Apple, Google, or Microsoft are trying to influence consumers' purchasing decisions in their favor through corresponding services. The car as "hardware" could thus become secondary, and the previous car manufacturers could be relegated to hardware suppliers. Naturally, they are trying to counteract this trend with their own platform offerings. Another strategy of car manufacturers is openness towards individual technology companies, such as Apple or

Google, regarding their operating systems and thus their ecosystems. Currently, it can be observed that German car manufacturers are entering into corresponding partnerships, such as that between Daimler and Google.

Once again, the pattern of hardware (in the form of cars and their manufacturers) versus software (in the form of infotainment and ecosystem providers) can be seen. An analogous example can be found in the field of machine control. The undisputed market leader in this area was and still is Siemens, among others with its S7 controllers. The company is trying to consolidate its leading market position by building its own platforms, MindSphere and Xcelerator. These offer additional services and further drive digitalization.

However, due to network effects and their sometimes high complexity, platforms are difficult for new market entrants to attack. The prerequisites are a large number of providers or cooperation partners and, of course, the services offered via the platform. This brings us to the "chicken-and-egg problem." In the early stages, platforms are often not very attractive. The goal is to quickly attract a large number of providers and partners. Just like digital marketplaces, they have marginal incremental costs. Successful platforms can thus build a quasi-monopoly and create significant barriers to entry for competitors. In addition, they will gradually claim an ever-larger share of the value creation for themselves.

Network effects lead to a corresponding amplification of one's own platform. Partner solutions not only increase the attractiveness of the respective platform, but also involve no effort or cost for the operator. The "problem" is that partners initially have little incentive to become part of the platform. However, once a critical mass is reached, rapid growth can often be observed. This can lead to the "winner takes it all" effect, i.e., a self-reinforcing effect. Companies can build a dominant market influence and are thus able to displace competitors or push them into niche markets, see [15].

Generating additional business through partnerships is certainly not a creation of Industry 4.0. What is new, however, is the combination of mostly digital services and platforms involving partners. This gives partners a completely different significance for the company itself. Of course, customers are also among the partners. Only through them is it possible, for example, to use data (see also Sect. 8.5.5) and, based on this, to generate new business models, as practiced by the American company Nest. The company supplies, among other things, intelligent heating devices and thermostats. However, customers only provide their data to the company if it creates added value for them. In this specific case, this is achieved by reducing energy costs by around 20%, see [15].

8.5.7 Digital Ecosystems

A digital ecosystem represents a further development of a digital business model—essentially the Champions League of the digital age. Such systems have already been introduced primarily from a technical perspective in Chap. 4. Examples of existing digital ecosystems include Apple, Google, and Microsoft. Apple has essentially created a world of hardware, software, and services, all offered through its own platform. Customers are provided with an "all-inclusive package." The combination of individual devices and services delivers additional value—if only because things simply work. Furthermore, the company collaborates with relevant partners. This results in significant customer value. Companies such as SAP or Siemens have also been working for years to build corresponding ecosystems.

The larger this ecosystem is, the more influential and market-defining these companies become. The size of the ecosystem is defined by the number of suppliers and customers (analogous to platforms) and the range of services offered. This sheer size is a crucial metric, especially at the beginning. In the B2B sector, the reputation of suppliers is another important criterion. This is particularly evident with the Siemens platforms MindSphere and Xcelerator. Well-known companies such as Atos or Accenture have been acquired as app suppliers, which in turn strengthens trust in the platform.

Digital ecosystems enable consistent thinking in user cycles, focusing directly on the tangible benefits and added value for the customer, see [13]. Closely linked to this is the generation of network effects. These arise when the value of a product or service changes with the number of customers, see [16].

All of the above examples are closed ecosystems, which is hardly surprising. This approach generates the best margins, as Apple's annual financial statements repeatedly demonstrate. The switching barriers for Apple users, for example, are significantly high, making it difficult to switch to another provider. Moreover, user satisfaction within these ecosystems is very high, which further reduces the motivation to switch.

However, there are also open systems, as demonstrated by the agricultural sector. Due to the circumstances (very few farmers operate exclusively with products from a single manufacturer), companies (such as Agco/Fendt) are compelled to adopt an open approach.

This overview shows that platforms and digital ecosystems are closely related in their mode of operation and their success criteria.

8.6 Transformation of Business Models

As with all other topics, the key to business models is to gain hands-on experience within the company and not to start out "too big." Courage and creativity are required here. The prerequisites for digital business models are suitable products, such as CPS, regardless of whether an evolutionary or disruptive approach is taken. The foundation for new business models has always been and will continue to be the core competence of the company itself. Combining this unique core competence with new technologies can lead to new products and business models.

Digital transformation affects individual parts of the business model or the entire business model, and thus the value chain. Transformation is also necessary for evolutionary approaches. Transformation involves changes in several dimensions:

- Customer dimension (customer segment and customer relationship),
- Value dimension (performance of the products),
- Partner dimension (partners and their relationships),
- Financial dimension (revenues, costs, value streams),

see [24]. The literature contains a variety of transformation models, see [24]. Fig. 8.3 presents a possible transformation.

In the four-stage model from Fig. 8.3, the first step is a critical assessment. This includes:

- current market position,
- new technologies,
- customer requirements,
- framework conditions (demographic change, compliance, etc.),
- employees and their know-how,
- the company's capacity for innovation and change (especially in management),
- knowledge management.

A key outcome is the identification of the company's own core competence, which must be sharpened and, if necessary, transformed. In the second phase, a corresponding strategy is derived based on a digital vision. The required roadmap should not exceed a period of two years. The roadmap should include new products and services, including

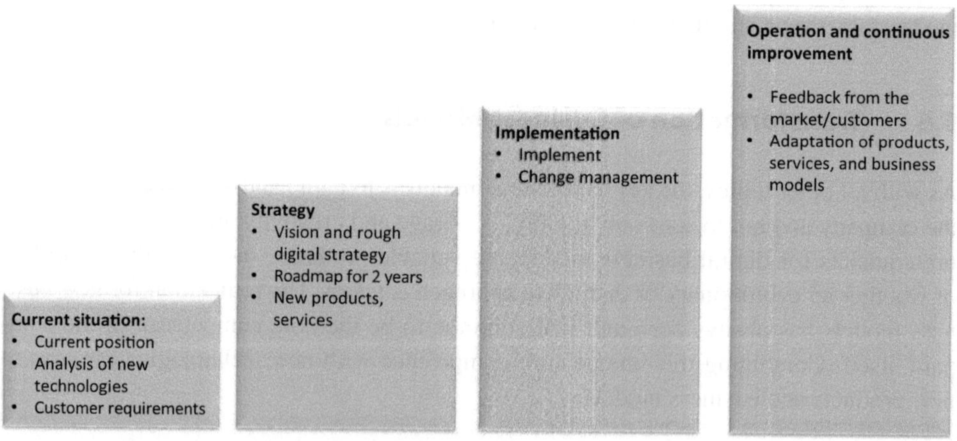

Fig. 8.3 Digital transformation of business models

business models. The goal should be: speed over scope—better to take the first small steps quickly than to launch a completely new digital product family in five years. The two-year period is chosen deliberately, as both technologies and the environment can change rapidly. Furthermore, the entire market is subject to strong dynamics, regardless of the industry considered.

The implementation phase is the most challenging part of the transformation. It is essential to "bring all employees along," that is, to engage in very active change management. This is where management is called upon; more on this in Chap. 9. Training and thus the qualification of the company's own employees is a central success factor here.

Based on customer feedback, the service and product offerings are adjusted. Continuous improvement should be a given. It is the key to success in conjunction with quality enhancement, see [9].

8.7 Challenges of Digital Business Models

Digitalization is not only a blessing but also a curse, presenting significant challenges for companies. In general, technologies pose a considerable risk if their relevance to the company's own business model is not recognized, not recognized in time, or not correctly assessed, and thus not adopted accordingly. The Gartner Hype Cycle, mentioned several times already, helps in assessing the technological maturity of individual technologies. The greatest potential—and thus, in part, the greatest risk—does not come from individual technologies, but from their interconnection, especially those we cannot yet imagine today.

This also includes the use of dynamic information. The value of information changes with time and place. The value of information thus depends on the overall context, see [13]. Future business models must take this fact into account and therefore require real-time data and information processing. For example, it is significant for a customer whether they are sitting in a car and receiving traffic information (such as about traffic jams) or on a train. In the latter case, such information is worthless, while in the former it is extremely helpful. New context-based business models can be generated from this, so-called location-based services, see [13].

8.7.1 Assessment of Technologies

The spread of digital technologies is often characterized by exponential growth. Unfortunately, people have difficulty grasping this characteristic and therefore tend to make serious misjudgments. The chessboard example illustrates this issue vividly. If you place 1 grain of rice on the first square, 2 grains on the next, 4 on the third, and so on, you end up with a total of 18,446,744,073,709,600,000 grains of rice on the last of the 64 squares—over 18 quintillion. The chip industry has experienced this kind of growth for decades, and currently, it is not the only one. Thus, at the beginning, growth is very slow, but over time it accelerates rapidly and gains tremendous momentum. This growth, along with the associated dangers and risks, is regularly underestimated. This can have serious consequences for one's own business model. At the same time, of course, new technologies also offer opportunities. Therefore, it is important to seize these opportunities and take a proactive approach to the topic.

This exponential phenomenon also applies to new ideas. In the future, it will therefore become significantly more difficult to predict innovations than it has been in the past. Moreover, there is an increasing combination/networking of a wide variety of technologies from different industries, resulting in new possibilities. This leads to explosive growth—not only of technologies but also of new business ideas. Consider the banking sector: in Germany alone, there are hundreds of start-ups, and around 12,000 worldwide, see [18]. If only 1% of these ideas succeed, it will fundamentally change the entire industry (1% is a realistic figure here). Currently, we are still in the first half of the chessboard from the example above. However, it gives an idea of where things might be heading. Therefore, specific defense strategies are needed. These will be discussed in the chapter after next.

However, as the experiences of recent years have shown, the impact of technologies on new business models should not be overestimated either. Change often occurs gradually (if we stick with the chessboard example, this would be in the lower half). Whether exponential growth occurs certainly depends on the specific case. As the above explanations have shown, such examples do exist—to the detriment of the affected industries and companies. Therefore, it is important to continue expanding one's own technical know-how in order to realistically and promptly assess potential risks.

8.7.2 Changed Customer Interaction

Another aspect is the so-called democratization of sales. Social media offers the opportunity to gain a very good overview of relevant products, including customer feedback. At the same time, it also provides sales with the opportunity to advertise (mostly indirectly). Sales costs are reduced, and there is a blending of sales and marketing. As a result, customer interaction is changing dramatically, especially in the B2C environment. Instead of a more or less pronounced one-way communication, dialogues are emerging, as can already be seen today (for example, in the form of chat functionalities). Multi-channel approaches are becoming increasingly relevant. These have already been mentioned several times and therefore require no further explanation.

8.7.3 Building Digital Trust

The new digital business models result in changed customer interactions and, among other things, challenges for customer loyalty. How can customer trust be built through digital media? The topic of "trust" is certainly a central issue in both B2B and B2C environments.

Trust in the digital world is created, among other things, through:

- Product and delivery quality,
- IT/cyber security,
- Data protection, the handling of data (for example, whether data is passed on to third parties), and data quality.

In addition, there are also the classic trust-building measures that exist in all business relationships.

It is not possible to achieve one hundred percent IT security. There is always a certain "residual risk." However, this residual risk must be minimized. The dangers do not only come from "outside," but increasingly from within. Bribery of employees, their careless handling of IT security requirements, and whistleblowers are just a few examples. As companies become more digitalized, the danger and risk of corresponding attacks increase. This is exacerbated by various political crises. Cyberattacks are increasingly state-directed or at least supported or tolerated. The topic of cyber security in its various forms has already been explained in previous chapters.

Of course, the technical protection of data is paramount. Approaches here include encryption of both data and communication. In addition, damage can be minimized by anonymizing the data. Another approach to minimizing damage is to avoid collecting data indiscriminately, but rather only those necessary for a service. This also conserves IT resources and ensures a lean and thus waste-free approach, see [9]. Another danger results not only from increasing networking but also from the ever-growing volume of collected data.

Data protection and cyber security are often only considered after a serious incident has occurred. There have already been ample reports of specific cases of damage, and the media are full of such events. For a long time, IT security was associated with antivirus software or firewalls. The headlines regarding targeted espionage attacks by intelligence agencies, alongside Industry 4.0, have led to much greater awareness of the issue. Here, too, it is important to sensitize and train one's own employees on the topic. With appropriate qualifications, employees can be an extremely effective firewall, but also a very significant risk (if the necessary awareness and qualification measures are lacking).

8.7.4 Handling Data and Data Quality

The introduction of new business models (and not only these, but all activities discussed here, from product development to production to new services and business models) often suffers from poor data quality within the company. This also includes the varying granularity and temporal resolution of the data (updates may occur every minute, hourly, weekly, etc.). Combining such data can therefore be problematic. The considerable data volumes must also be processed and stored accordingly. The necessary infrastructure for this must be built in a timely manner. Along with building the infrastructure, it is also important to develop know-how in handling the data, from data aggregation and analytics to AI applications. Overall, considerable effort is required for system integration—including towards suppliers. The required vertical and horizontal integration has already been discussed in Chap. 5.

Furthermore, the real-time capability that is often required (not only of the data but also of the participating systems) is often not available. In addition to the data and systems, the entire internal infrastructure must also be aligned with the new requirements.

8.7.5 Infrastructure

For “always on”—including for CPS—a fast internet connection is simply essential. Unfortunately, Germany still lags significantly behind the global average in terms of internet expansion. In 2017, Germany ranked 25th with an average speed of 15.3 Mbit/s. By April 2023, this had improved to approximately 80 Mbit/s (download) and around 29 Mbit/s (upload), see [40]. The leaders here are Singapore and China with over 220 Mbit/s [3]. The situation is quite different in the mobile network sector. Here, Germany ranks second behind the United Kingdom with 24.1 Mbit/s, see [4]. Nevertheless, making phone calls on German highways is still often a challenge. Compared to 2017, the situation in Germany has significantly worsened. Currently, providers in Germany rank 16th in a European comparison, see [7].

The picture is more positive when it comes to the innovation index in international comparison. Here, Germany ranks fourth, one of the top positions, see [12].

A similarly poor picture as with internet connectivity emerges with the so-called digitalization index, which is compiled every two years by, among others, the Fraunhofer Institute for Systems and Innovation Research (ISI). This index examines and analyzes the state of digitalization in Germany at the federal state level. Here, Germany ranks 17th, placing it in the middle of the field, see [12]. However, the situation has improved somewhat compared to 2019, see [41]. Thus, there is a significant need for government action. More on this in Chap. 10. However, this index does not provide any information about the degree of digitalization of German industrial companies. In the private sector, the topic is taken very seriously, with corresponding implementation programs. In general, German industry can be given a much better report card than government agencies.

8.7.6 Dealing with Legacy Systems

Unfortunately, the world is not made up entirely of new machines, equipment, and infrastructure with the corresponding networking capabilities. In the transport sector (aviation, rail, shipping), equipment has correspondingly long service lives due to high acquisition costs. For example, locomotives are only replaced after 15 to 30 years, see [14]. The fleet of aircraft is also only replaced with the introduction of new models, which also results in service lives of over 20 years, and sometimes even up to 40 years. In the private household sector, refrigerators and washing machines also have similarly long lifespans. This creates a need for migration. For machines and equipment, this can sometimes be achieved by modernizing the control systems. It is estimated that around 40–50% of equipment in the transport sector would need to be replaced, see [14]. This would require corresponding investments. The realistic scenario is that, for the next few years, there will certainly be a coexistence of non-upgradable equipment alongside new or modernized machinery. This leads to a further increase in complexity. In agriculture, at least partial retrofitting is carried out using appropriate boxes, such as those offered by the Berlin start-up 365Farmnet, see [11]. Smart sensors help to generate data that is not otherwise available. There are also numerous secondary approaches to obtain information about the condition of machines, such as by monitoring power consumption.

As is often the case, legacy systems also present new opportunities. More and more manufacturers are buying back their old equipment and modernizing it, for example by installing new control systems, sensors, graphics cards, etc. Examples include Siemens Healthineers Refurbished Systems and Bosch. These "refurbished" devices are then primarily sold in emerging markets, including a service package. In addition to opening up new sales markets, this also extends the respective product life cycle. The focus here is on durable products, such as those from the medical sector, see [31].

8.7.7 Organizational Aspects

New business models often also require organizational changes within companies. This topic is addressed, among other things, in Chap. 9. The complexity and effort involved should not be underestimated by companies. Active change management is essential for this. It is important to involve your own employees at an early stage. Experience has shown that this can significantly increase acceptance. Employees do not want to be changed; they want to actively participate in the change process.

Change management can be supported by actively training your own employees. Above all, the approval of the works council is likely to be assured for this. This training is also necessary because the transformation to digital business models entails a significant shift in the personal knowledge and know-how of every employee.

Currently, there is only limited experience with long-term successful digital business models. In many cases, attempts are made to draw parallels with previous upheavals, such as the advent of the PC, see [14]. Unfortunately, these comparisons are only of limited help. The experiences from lean transformation are much more useful here. Companies must either gain experience themselves or exchange ideas through networks. There are a variety of opportunities for this, some of which are promoted by professional associations or universities.

In many cases, introducing new business models in a company can only be managed with considerable organizational effort. This is where start-ups come into play. Actively engaging with and supporting them can also be profitable for “large” companies. Osram, for example, supports start-ups not only financially but also with many organizational issues (access to international markets, IT-related topics, legal issues, etc.).

Unfortunately, only small amounts of venture capital are available in Germany, at least when compared to the sums in the USA. The gradual difference within Germany, however, is understandable. Berlin is still considered the start-up capital, followed by Stuttgart, Munich, and Hamburg, see [10].

In addition, the transformation of a company’s own business model also affects relationships with its suppliers, especially when it comes to disruptive/revolutionary approaches.

8.7.8 Political Aspects

Regardless of Germany’s progress in digitalization and thus the local approach, there is an urgent need for a European digital single market with uniform rules. Digital business models in particular depend on a large, unified market, if only to be able to offer a comparable homogeneous economic area to America or China. Digital business models are characterized, among other things, by the fact that they can easily be exported across national borders. The European Commission has taken a step in the right direction with its strategy for a digital European single market, see [26]. However, there is still a need to catch up in its implementation (such as Gaia-X). The handling of potential regulation of AI applications also requires a much more realistic assessment of opportunities and risks.

Furthermore, it must unfortunately be noted that regulatory barriers in Germany are significantly higher than, for example, in the USA. A concrete example is autonomous driving. This is handled much more easily for companies not only in the USA but also in Sweden. On the other hand, regulatory requirements in the area of data protection in Europe and Germany are much more favorable for digitalization than, for example, in the USA, see [20]. Once again, there are both advantages and disadvantages. These regulatory barriers mean that companies are driving corresponding innovations outside Germany.

References

1. Ömer Atiker, In einem Jahr digital: Das Praxisbuch für die digitale Transformation Ihres Unternehmens, Weley-VCH Verlag, März 2017.
2. Werner Beutnagel, Landpartie, automotiveIT 08/08 2017, Seite 38–39.
3. https://www.internetgeschwindigkeit.net/search/internetgeschwindigkeit+weltweit/#:~:text=%C3%9Cberraschend%20ist%2C%20dass%20L%C3%A4nder%20wie%20Tschechien%20oder%20Rum%C3%A4nien,Spitzenpl%C3%A4tzen.%20Die%20globale%20durchschnittsgeschwindigkeit%20betr%C3%A4gt%20derzeit%205%2C1%20Mpbs. Aufgerufen am 17.06.2023.

4. Mathias Brandt, Deutsches Mobilfunknetz ist spitze, statista, 12.06.2017, https://de.statista.com/infografik/9770/durchschnittliche-mobile-internetgeschwindigkeit, aufgerufen am 30.08.2017.
5. Birgit Buchholz, Jan-Peter Ferdinand, Jan-Heinrich Gieschen, Uwe Seidel, Digitalisierung industrieller Wertschöpfung- Transformationsansätze für KMU, Begleitforschung AUTONOMIK für Industrie 4.0, April 2017.
6. Volkhard Emmrich, Mathias Döbele, Thomas Bauernhansl, Dominik Paulus-Rohmer, Anja Schatz, Markus Weskamp, Geschäftsmodell-Innovation durch Industrie 4.0, Studie, März 2015, https://www.wieselhuber.de/migrate/attachments/Geschaeftsmodell_Industrie40-Studie_Wieselhuber.pdf, Aufgerufen am 22.08.2017.
7. Danny Fischer, Im EU-Vergleich landen deutsche Handynetze weit hinten, 8. Mai 2020 SmartDroid, https://www.smartdroid.de/im-eu-vergleich-landen-deutsche-handynetze-weit-hinten/ Aufgerufen am 17.06.2023.
8. Santiago Gallino et al: „Need für Speed: The impact of In-Process Delays on Customer Behavior in Online Retail“, Operations Research, im Druck, Harvard Business Manager, 07/2023.
9. Michael Heidermann, Zeppelin – Neue Services und Lösungen für die Bauwirtschaft, Vortrag, Horst Wildemann (Hrsg.) Münchner Management Kolloquium, 14.3. – 15.3. 2017, Tagungsband Seite 259–272.
10. Markus Heyn, Servicelandschaften im Umbruch: Neue Mobility Services revolutionieren den Automobil-Servicemarkt, in Claudia Linnhoff-Popien, Michale Zaddach, Andreas Grahl (Hrsg.), Marktplätze im Umbruch: Digitale Strategien für Services im Mobilen Internet, Springer Vieweg, April 2015, Seite 49–56.
11. Masaaki Imai, KAIZEN – Der Schlüssel zum Erfolg im Wettbewerb, Econ-Verlag, 1986.
12. W. Chan Kim, Rene'e Mauborgne, Beyond Disruption: Innovate and Achieve Growth without Displacing Indutries, Companies, or Jobs, Harward Business Review Press, 2. Mai 2023.
13. Tobias Kollmann, Christoph Stöckmann, Simon Hensellek, Julia Kensbock, Deutscher Startup Monitor 2016, Bundesverband Deutsche Startups e. V., 2016.
14. Philipp Krapp, Josef El-Rayes, Michael Zettel, Digitalisierung – Konjunkturmotor in der Krise, Bundesministerium Digitalisierung und Wirtschaftsstandort, Juni 2020,
15. Nils Kreimeier, Twitter für Trecker, Capital, 08/2017, Seite 90–94.
16. Stefan Krempl, Innovationsstudie: Deutschland hinkt bei Digitalisierung und Glasfaser hinterher, Heise Online, 24.07.2017.
17. Ralf T. Kreutzer, Karl-Heinz Land, Digitaler Darwinismus: Der stille Angriff auf Ihr Geschäftsmodell und Ihre Marke, Springer Gabler, Februar 2016.
18. J. Manyika, M. Chui, P. Bisson, The Internet of Things: Mapping the value beyond the hype, McKinsey Global Institute, https://www.mckinsey.de/files/unlocking_the_potential_of_the_internet_of_things_full_report.pdf, aufgerufen am 23.08.2017.
19. Thomas Matyssek, Geschäftsmodelle im Internet der Dinge, in Daniel Schallmo (Hrsg.) Digitale Transformation von Geschäftsmodellen, Springer Gabler, November 2017, Seite 159–177.
20. Kurt Matzler, Frank Bailom, Stefphan Friedrich von den Eichen, Markus Anschober, Digitale Disruption: Wie Sie Ihr Unternehmen auf das Digitale Zeitalter vorbereiten, Verlag Fran Vahlen, 2016.
21. Andrew McAfee, Erik Brynjolfsson, Die große Abkopplung, Harvard Business Manager, September 2015, Seite 32–41.
22. Jürgen Meffert, Heribert Meffert, Eins oder null, Econ Verlag, April 2017.
23. Martin Mehringer, Der Nerd und sein Bodyguard, manager magazin, September 2017, Seite 59–62.

24. Simon C. Müller, Markus Böhm, Helmut Krcmar, Isabell M. Welpe, Machbarkeitsstudie: Geschäftsmodelle in der digitalen Wirtschaft, Studien zum deutschen Innovationssystem Nr. 12-2016, Technische Universität München, Februar 2016.
26. Janessa Rivera, Rob von der Meulen, Gartner's 2014 Hype Cycle for Emerging Technologies Maps the Journey to Digital Business, http://www.gartner.com/newsroom/id/2819918, aufgerufen am 11.09.2017.
27. Alix Sauer, Michael Stark, Digitalisierung – Wir müssen die Treiber sein, Aktiv, Juli 2017, Seite 8–9.
28. Daniel R.A. Schallmo, Jetzt digital transformieren: So gelingt die Digitale Transformation Ihres Geschäftsmodells (essential), Springer Gabler, August 2016.
29. August-Wilhelm Scheer, Nutzentreiber der Digitalisierung – Ein systematischer Ansatz zur Entwicklung disruptiver digitaler Geschäftsmodelle, Informatik Spektrum, Band 39, Heft 4, August 2016, Seite 275–289.
30. Dieter Schweer, Jan Christian Sahl, Die digitale Transformation der Industrie – wie Deutschland profitiert, in Ferri Abolhassan (Hrsg.), Was treibt die Digitalisierung, Springer Gabler, Dezember 2015.
31. Thomas Steinmann, Markenmoment, Capital, 08/2017, Seite 16.
32. Gunter Trojandt, Deutschland digital – Algorithmus-Ärzte, Capital, 08/2017, Seite 14.
33. Frank Weber, Joachim Berendt, Robuste Unternehmen: Krisenfest in Zeiten des Umbruchs, Springer Gabler, Mai 2017.
34. Horst Wildemann, Produktivität durch Industrie 4.0, 24. Münchner Management Kolloquium, 14./15. März 2017.
35. Horst Wildemann, Neue Geschäftsfelder. Geschäftsmodelle, Technologien: Müssen sich Unternehmen neu definieren? TCW-report Nr. 69, Verlag TCW Transfer-Centrum, 2017.
36. Wie sich Unternehmen neu erfinden: Digitalisierungsprojekte bei Grohe, Hella und Kiekert, CIO, 30.08.2017.
37. https://medium.com/world-of-iot/73-lessons-to-learn-from-the-failure-of-ges-iot-platform-predix-3b3d20eccd42, aufgerufen am 11.06.2023.
38. https://markteinblicke.de/76502/2020/10/walmart-vs-amazon-kampf-der-giganten-um-den-e-commerce-thron, aufgerufen am 11.06.2023.
39. https://www.carsharing-news.de/carsharing-anbieter, aufgerufen am 17.06.2023.
40. https://de.statista.com/statistik/daten/studie/416534/umfrage/durchschnittliche-internet-geschwindigkeit-in-deutschland.
41. https://www.fokus.fraunhofer.de/de/news/fokus/dps/D-Index_2021_05, aufgerufen am 18.06.2023.

New Companies and Leaders 9

Regarding digitalization and Industry 4.0, one might get the impression that such a transformation is solely a technical challenge. However, this admittedly engineering-focused perspective is unfortunately a misconception. While digital technologies do change companies, taking advantage of the new opportunities they offer and meeting market challenges requires much more—namely, the transformation of the entire organizational model and the company's own processes. In order to remain competitive, companies must be agile, efficient, innovative, and adaptable (see also the introductory chapters)—in other words, robust, see [29], [31]. For many companies, this means profound change. Achieving this requires a high degree of knowledge management is required. These changes therefore affect all levels and areas, including the organizational model, managers and executives, and ultimately every single employee.

Likewise, long-term strategies must also be abandoned. Volatile markets and new disruptive technologies no longer allow for multi-year strategies. Thus, the journey becomes the goal. This also applies to the annual budget rounds that take place in companies. Often, the painstakingly determined budgets are already partially outdated by the time they are approved. This results from an increasingly dynamic market environment. This is generally true for all industries, though with varying degrees depending on the sector. However, this dynamic environment should not prevent leadership from developing a corporate vision. The strategy derived from this vision and the concrete projects should be, as a rule of thumb, be planned for about two years (as already mentioned). A vision is designed for the long term and should therefore remain in place. The underlying strategy, on the other hand, must be developed dynamically. There is little point in engaging in lengthy strategy discussions. It is more important to start with concrete implementation and gain experience within the company. Based on these experiences, an appropriate strategy can then be developed. This must be continuously adapted to changing market conditions.

W. Huber, *Industry 4.0 Compact—How Technologies Are Transforming Our Economy and Our Companies*, https://doi.org/10.1007/978-3-658-51029-9_9

At the same time, the Fourth Industrial Revolution threatens not only simple tasks and thus jobs (primarily in production environments), but increasingly also those of academics and activities in indirect areas. Digital technologies are to human intellectual capabilities what the steam engine and similar technologies were to human muscle power—namely, a significant increase in specific performance, see [19]. The rapid development of AI, especially LLMs (Large Language Models) such as ChatGPT, will further accelerate this dynamic. To what extent such productivity gains will lead to actual job losses in companies remains to be seen. Even with the lean transformation, there were considerable productivity increases and thus concerns, especially from works councils. At that time, job losses were feared. The resources saved were often used for further optimizations or to increase revenue. As a result, the work content of employees changed. Only those employees who were unwilling or unable to adapt were affected by job losses. In digital transformation as well, the resources saved can be used for innovation and revenue growth. Thus, lessons can be learned from the experiences of the lean transformation already carried out. These experiences give hope that no massive job losses are to be expected. Moreover, there is a significant shortage of skilled workers in virtually all industrialized countries. Digital transformation can help address or at least mitigate this shortage.

For companies, the situation arises that knowledge accumulated over decades is being massively devalued by smart products and new technologies such as machine learning. This often leads to massive resistance to impending and necessary changes, especially when business is going well ("Don't touch a running system" mentality among managers). People and organizations must leave familiar comfort zones, see [15], [29]. As a result, companies are often only willing to change when crises have already occurred. By then, however, it is often too late. Therefore, it is essential to continuously and critically—but also constructively—question not only one's own products and business models, but also the organizational model.

The development of a smart company, that is, the digital transformation of the organization, can be divided into several, sometimes overlapping, subject areas:

- Transformation of the company primarily from an organizational perspective,
- Change in management—keyword Management 4.0,
- new employees with appropriate digital know-how.

The individual subject areas will be examined in more detail below.

9.1 Company Transformation

Markets and customer needs are currently changing rapidly, and the pace will continue to increase. Private life has also undergone significant changes due to digitalization, which in turn influences expectations for companies, products, and services. On the other hand, corporate structures and decision-making processes have not changed significantly since the 1960s. As a result, a large proportion of companies are often too slow for the digital age. Rigid budgets are often already outdated by the time they are approved (as already mentioned). Here, one can take a cue from the approach of venture capitalists. This companies only release budgets for a certain scope at a time. Depending on the achievement of objectives, further funds are then released.

Companies must therefore quickly adapt and respond to the changes already mentioned several times. New management, decision-making, and leadership processes need to be established. Through diversity, companies have a significantly higher probability of growth (45%) and of conquering new markets (70%), see [18].

However, when it comes to technological change, it is important to note that its speed is not an absolute measure (as has been mentioned several times). Each industry has its own dynamics. In aerospace, the wheels of change turn much more slowly due to very long product life cycles and very high compliance requirements than, for example, in the fast-moving mobile phone industry. Every company also has its own individual transformation speed. It is important to know and take this into account. This also includes knowledge of the company's own values and culture. If these are neglected, any transformation becomes a shot in the dark.

Anyone who wants to radically rebuild their way of working and processes needs a corresponding cultural foundation. It is not without reason that Peter Drucker coined the phrase "culture eats strategy for breakfast," which can be adapted to "culture eats digital for breakfast," see [26]. Digital transformation represents a cultural change within the company.

Regardless, proactive action is required, meaning that companies must shape the market and thus remain or become innovation leaders. This also involves changing and adapting business models, products, and their own production, including the supplier/partner network—in other words, the digital supply chain.

Conversely, reasons for the failure of companies become apparent:

- Defensive thinking: This often arises among market leaders. Employees then begin to think defensively and thus conservatively. Innovations and changes are viewed much more critically. A good example of this is Agfa and Kodak (as already mentioned several times).
- Inflexible corporate systems and structures: These structures often arise in growing companies. Here, energy and focus shift from innovation to exclusive improvement. As a result, innovation is greatly neglected. This shift has a positive effect on efficiency, but at the same time a negative effect on the ability to innovate and adapt,

see among others [9]. It is important to learn from these mistakes.

9.1.1 Organizational Aspects

Smart products not only impact the underlying business model, but also require cross-departmental collaboration, from development through to service. This leads to the breaking up of traditional structures and departmental boundaries, resulting in new organizational models as well, see [33] and Fig. 9.1. The insight that cross-departmental collaboration is necessary is not new; it has long been known from the Lean world, see [12]. Therefore, before any digital transformation, a Lean transformation towards streamlined and stable processes and corporate and organizational models should take place. Technologically, cross-departmental collaboration requires both vertical and horizontal integration, as partners also need to be involved. While this cross-departmental collaboration may sound simple, in practice it presents significant challenges and many hurdles must be overcome. These changes must be initiated by top management.

To meet these challenges, companies must set out on the path to becoming a robust enterprise, as mentioned at the beginning of the chapter.

To drive innovation within a company, all management processes must be questioned and, in most cases, restructured. Bold thinking and radical action need to be encouraged. This includes flat hierarchies and the aforementioned decision-making processes, as well as cross-departmental collaboration and working in project teams rather than rigid line organizations. Employee engagement is a key factor and must be fostered.

As a consequence of new agility, decisions must be made quickly, at the right time and in the right place. This results in a shift of decision-making to lower management levels. This promotes decision-making autonomy and leads to decentralized structures.

Adaptability is more important today than ever. This is also reflected in the term Digital Darwinism. The more "organized" and strictly "managed" companies are, the less adaptable they become. One could also say that hyper-rational managers produce

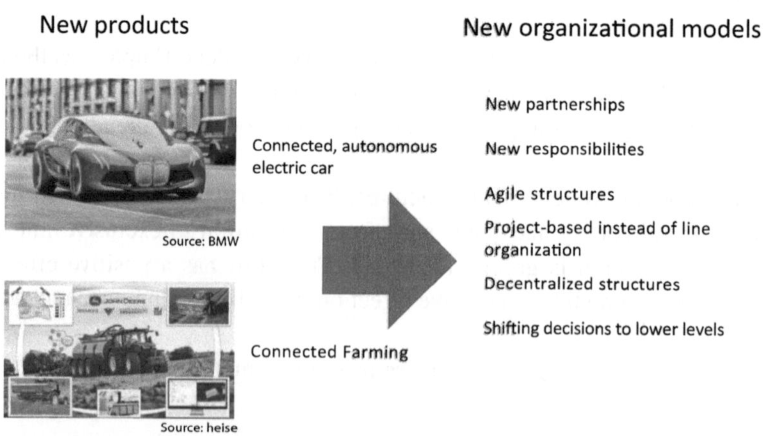

Fig. 9.1 New products require new organizational forms

hyper-boring products, see [9]. Adaptable companies must continually rethink their core business so that they can recognize and seize new growth opportunities at any time. In this way, companies are constantly reinventing themselves and opening up new markets. This is also referred to as constructive destruction (again, a term from the Lean world). Companies must always be ready to abandon existing (especially unsuccessful or less successful) business. The goal is to learn early, quickly, and cheaply. These are the rules for companies that do not want to be overtaken by competitors and change, see [9]. However, this also requires employees who are willing and able to adapt and change. Without them, it is impossible to build an adaptable company. Unfortunately, many people only change when they have to. More on this in Sect. 9.2.

Just as with Lean, digitalization also requires a change in behavior. Employees must be made aware not only of innovation, but also of continuous digital improvement. This is a task for both the company and management. Products and services must be continuously improved. Companies must therefore learn to align products and processes much more consistently with the customer and thus build a customer-centric organization.

To master all these challenges, certain roles and their understanding must be specifically redefined, see Fig. 9.2. It should come as no surprise that the CIO (i.e., the Head of IT) will gain in importance. The traditional assignment under the CFO is no longer appropriate. IT is becoming a strategic factor and must therefore be positioned alongside the classic board areas, with a significantly higher budget. IT costs should no longer be seen as a cost factor, but as a source of innovation. Close coordination with product development and other departments such as production is also necessary. Where product-specific software developers are located is secondary—the important thing is that they exist at all. IT itself must also change. Technology must increasingly be seen in connection with business challenges. IT staff must develop a certain level of business knowledge in order to discuss and collaborate with their colleagues from the respective departments "at eye level." Of course, this also applies in the other direction, i.e., for the business departments. This also helps dissolve rigid departmental boundaries. Regardless

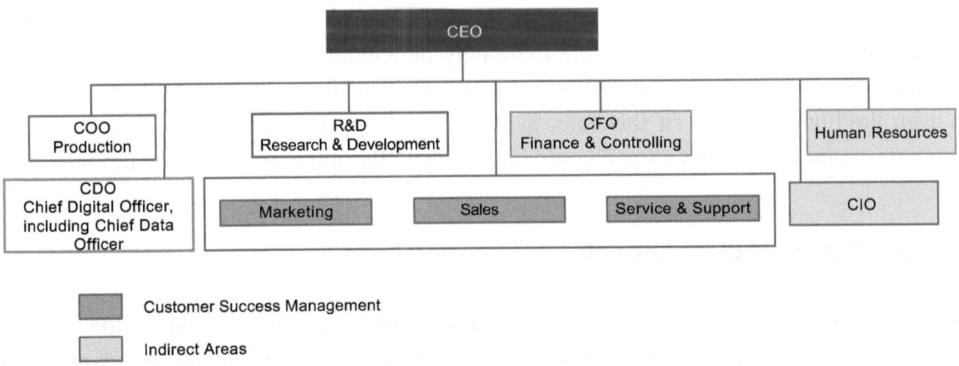

Fig. 9.2 New organizational structure

of the new role of IT, it must be oriented towards the needs of the business departments, acting as their service provider. Digital transformation must primarily take place in and be led by the business departments, see [7]. This is where value must be generated. Conversely, the business departments must build up a considerable level of digital competence. In return, IT acts as an internal consultant for the business departments and as the technical "enabler."

IT requires a significant increase in cyber security experts, among other things, to secure the new smart products and their communication with each other. The technical setup of a CPS infrastructure discussed in previous chapters also falls within the responsibilities of IT, while the business departments must provide the content and requirements.

Parallel to the CIO, there is the CDO—the Chief Digital Officer. It is important that the CDO sees themselves as a business contact. While they certainly have an IT background, they see themselves more as the "innovator" within the company and are therefore organizationally located within the business department. In a sense, they are the "client" for IT. Thus, it is clear that the CDO has their own budget. They must work closely with the respective business departments and IT to create and be responsible for the relevant business cases. The CDO is tasked with overcoming silo thinking, as their role is inherently cross-departmental. To achieve this, they need the appropriate competencies; otherwise, they will quickly become a "paper tiger" and be slowed down by the business departments. The CDO team also includes big data experts. Through them, added value is created for the respective business departments, which increases the acceptance of the entire team within the company. Furthermore, the CDO is also responsible for the increasingly important partner and network management and, if necessary, for building a company-owned digital ecosystem. Exchanging with other companies increases one's own speed, helps avoid mistakes, reduces costs, and also fosters innovation. The CDO is therefore responsible for the entire digital transformation of the company. In larger companies, it makes sense for the CDO team to also include dedicated transformation personnel. It should be clear that this is not a side job. With such far-reaching changes, it should come as no surprise that the CDO may sometimes make unpopular decisions. Here, they absolutely need the backing of top management. The description makes it clear that combining the roles of CIO and CDO is not sensible. The responsibilities of both positions are extremely demanding and sometimes even contradictory. The larger the company, the more pronounced the role of the CDO, see [10]. Given the high demands of this role, it is not surprising that, despite its importance, it remains a rare species in German companies, especially in the mid-sized sector.

9.1.2 Technological Aspects

In order to successfully implement digital transformation within an organization, it must be technologically supported and underpinned. Internal processes need to be adapted and changed. These processes must become lean in the digital sense, meaning, among other things, low-paper or ideally paperless. This requirement is not entirely new. For example, paperless manufacturing is based on a modern production control system (currently an MES—Manufacturing Execution System, and in the future a Cyber Physical Production System, or CPPS for short); otherwise, paper-based processes risk leading to waste, lack of transparency, and thus efficiency losses, including slow and delayed responses and decisions to events. Paperless processes are also the foundation for collecting company data in a big data system. Only through end-to-end digitalization of all processes can the many listed added values and benefits be realized and implemented. Paperless manufacturing is much more than a single system. This includes a digital qualification matrix, shift planning, integrated work instructions, mobile solutions for maintenance, to name just a few. These have a significant impact on the scope of work and also on transparency. This high level of transparency must be handled constructively.

Corporate communication via email has long been considered a source of inefficiency. It is estimated that around 204 million emails are sent worldwide every minute, see [24]. Collaboration solutions provide significantly improved information exchange. This trend was further reinforced by the coronavirus pandemic and is now largely irreversible. Shop floor meetings in the classic lean sense can also significantly reduce the flood of emails. The combination of collaboration solutions with lean approaches delivers further improvements. The goal here is to transform existing lean concepts such as shop floor boards into the digital age without losing the benefits that have been achieved. Here, too, the aim must be to become lean in the digital sense. There are various solutions for digital shop floor boards available on the market, like from Bosch (Active Board). This not only enables the digitalization of collaboration and communication, but also decouples them from location, i.e., physical presence. The shop floor meeting and its associated board can thus be expanded into a general, company-wide collaboration board, even across site boundaries. Actions can be derived from automatically generated key figures and monitored using tools.

Collaboration solutions enable employees to avoid spending their entire working hours in the office. In other words, they can also work productively while traveling or from home. As already mentioned, this trend was further reinforced by the coronavirus pandemic. The practicality of this approach has also been demonstrated. It is therefore not surprising that employees expect corresponding home office arrangements. At the same time, valuable office space is saved without introducing open-plan office concepts. According to an American meta-study, these actually inhibit employees' willingness to innovate and their productivity, see [1]. The "compression" of office space is therefore completely unnecessary and counterproductive. Overall, these individual measures promote the separation of work content and work location and at the same time increase work-life balance at lower cost.

Collaboration solutions can also help save on business trips or avoid the annoying morning traffic jams. All communication can be digitalized and applications can be much better integrated into digital workflows. Currently, office products are used for many tasks. Afterwards, distribution is done by email and documents are printed out.

This creates a break in the digital process chain. Obtaining information is already difficult in traditional office structures; in the absence of employees, they have to rely on emails, making information retrieval a very tedious task. Employees who travel a lot often need information quickly. Collaboration solutions provide an answer here. This way, employees have access to all, and above all, the most up-to-date information from anywhere at any time. Decision-making processes are significantly accelerated and work becomes more efficient. They also promote open communication within a team, which can help prevent misunderstandings and ensure that information flows more quickly, see [21]. It is also important to ensure good transparency. This is the basis for efficient communication. Every team member must know what the others are working on. Communication in the digital age still consists of listening, understanding, and speaking.

In web meetings, information can be exchanged easily and in a timely manner. Teams can focus primarily on solving problems rather than exchanging data and information (which has already taken place in advance). According to McKinsey, employees spend about 20% of their time obtaining information and an additional 28% reading emails, leaving only 39% for their actual work. Thus, with optimal work organization, a productivity increase of 20–25% is quite possible for so-called knowledge workers, see [5].

As a consequence, mobility will continue to increase; currently, nearly 40% of employees are mobile, and by 2025 this will rise to 42.5%, see [21]. This opens up new opportunities for companies outside metropolitan areas to attract employees and increase their own attractiveness. In addition, the carbon footprint improves as commutes to the workplace can be reduced.

Machine learning is being adopted not only in direct areas but also in indirect areas, making employees more productive. In sales, productivity can be increased by about 9% according to McKinsey, and in traditional office areas between 16 and 40%. Energy management can also be significantly improved—by about 12–21%, see [17].

Mobile solutions are only in use in one out of every seven departments in Germany. The low penetration is particularly pronounced in areas close to manufacturing, see [4]. In contrast, a global study by Strategy Analytics found that 38.8% of employees use mobile solutions, and by 2022 this share will rise to over 40%, see [16]. These seemingly contradictory statements result from a much higher use of mobile solutions in Africa and Asia due to a significantly younger workforce, see [16]. Germany is therefore somewhat lagging behind the general trend. For 2027, the market volume for such solutions is expected to be just under US$10 billion, up from just under US$5 billion in 2020, see [35].

9.1.3 The New Role of the HR Department

The transformation of companies inevitably leads to changes in personnel requirements and demands on the HR department. Employees with IT skills will make up a significantly higher proportion of the workforce in the future than is the case today. In general, it is important to have the right employees with the right skills on board at the right time for the company. This requires a personnel strategy and staffing needs derived from the corporate strategy. This is not easy given the current shortage of skilled workers. However, the shortage of skilled workers should also be questioned. Hiring employees over the age of 50 or 60 should not be taboo. Creative ideas are needed here, and potential employees should, if necessary, be further qualified internally.

The HR department must also develop significantly in terms of expertise. It must be the driver for change management towards a smart organization. Distributed work, home office, continuous further training, and new career models are just some of the challenges. These approaches must be flexible and individually adapted to the respective life situations of employees. The career model must be much more permeable and offer more qualification and thus advancement opportunities and remuneration for skilled workers. This also includes greater involvement and motivation of employees in and for innovation. Otherwise, companies risk losing the race for the best "brains." This also means working on their own image. Companies like Audi, BMW, or Google have less trouble attracting highly qualified new employees than, for example, medium-sized companies. Thus, the shortage of skilled workers is at least partially put into perspective. Therefore, companies need to work on their own image. The fact that there is still room for improvement is evident, for example, in the still often very lengthy application processes for new employees. Here, speed and flexibility are required. Applicants generally do not fit job postings "tailor-made," and they do not wait indefinitely for a response. They simply go to the competition.

In general, the approaches already mentioned should be used to retain employees in and with the company in the long term. These are therefore less capital-intensive measures. These measures serve more to increase flexibility on both sides. Higher salaries have long ceased to be a motivator for employees, at least not in the long term.

The HR department must also prepare managers for their new tasks. Developing appropriate training programs is therefore far from sufficient. Leading distributed teams using digital tools is much more complex than existing approaches. This is further complicated by multicultural aspects. We will discuss this in more detail later in this chapter.

9.1.4 Transformation Process

The goal of the digital transformation of internal processes is to improve efficiency and thus company performance, while simultaneously building the necessary competencies. According to an IDC study from 2017, the majority of German companies (exactly 64%) have launched corresponding projects, and only 16% have reached a level of maturity that allows them to reap the associated benefits. In contrast, around 20% are still taking a wait-and-see approach, see [11]. According to the German Ministry for Economic Affairs, in 2020, 74% of German industrial companies had a digitalization strategy and 70% were investing in digital training and further education for their employees, see [36]. Thus, a slight improvement can be observed over the past five years.

One challenge is that employees and management are at different stages of the change process. While the boss or management is usually already in the implementation phase, employees are still at the beginning of the process. This is partly due to the different timing of their involvement and the degree to which they are engaged with the topic.

The transformation process begins with a vision. Every new vision means change, and every change begins in the mind. This change must not only originate from top management, but must also be supported by the leadership team, and there must be a shared vision or one that is developed together. It is important to note that this is not a classic project with a defined end and thus a fixed duration, but rather a continuous process that will certainly extend over years. More precisely, it will last until the 5th Industrial Revolution, whenever that may be. Furthermore, it is not the degree of digitalization that is decisive, but rather how it creates maximum value for the company and its customers. Thus, here too, the customer and customer benefit are the "yardstick." Moreover, the focus should not be on implementing what is technically possible, but rather on concentrating on economically sensible measures.

Companies should therefore, as with lean transformation, be strictly oriented toward the customer and their needs. Consequently, the processes with the highest customer benefit should be addressed first. Innovation and product development processes must undergo significant change. It is not enough to believe that new products and services can be created with old processes. While this may work temporarily, at the latest in the medium term, this approach is doomed to fail.

As with lean transformation, open communication and early involvement of employees and the works council are essential. There must also be space for employees to communicate with each other. Communication is generally extremely important for cross-functional collaboration (regardless of whether this is tool-based or verbal). There is often the misconception that open office structures promote communication. In the aforementioned meta-study, the exact opposite was proven, see [1].

The transformation of an organization should be sustainable. It is not enough to hire new managers and define new job descriptions and titles. Sustainable organizational change takes time. Therefore, it is advisable not to approach this topic too radically. All employees in the company, including the works council, must be brought along on the digital journey. Thus, the transformation should take place in individual steps. The teams required for this should have a mixed structure of experienced colleagues and younger employees. Leaving the topic solely to "Generation Y or Z" would be grossly negligent. Furthermore, direct employees in production must also be included. Training opportunities and facilities must also be created for this group of employees, which often proves to be the more difficult part compared to indirect areas. These employees generally do not have their own IT accounts or email addresses. Therefore, alternative approaches must be developed here. Cloud-based learning platforms in combination with bring-your-own-device can be a solution, see also Fig. 9.3. In many cases, a change in thinking is also taking place. Direct employees are increasingly being given corresponding accounts in

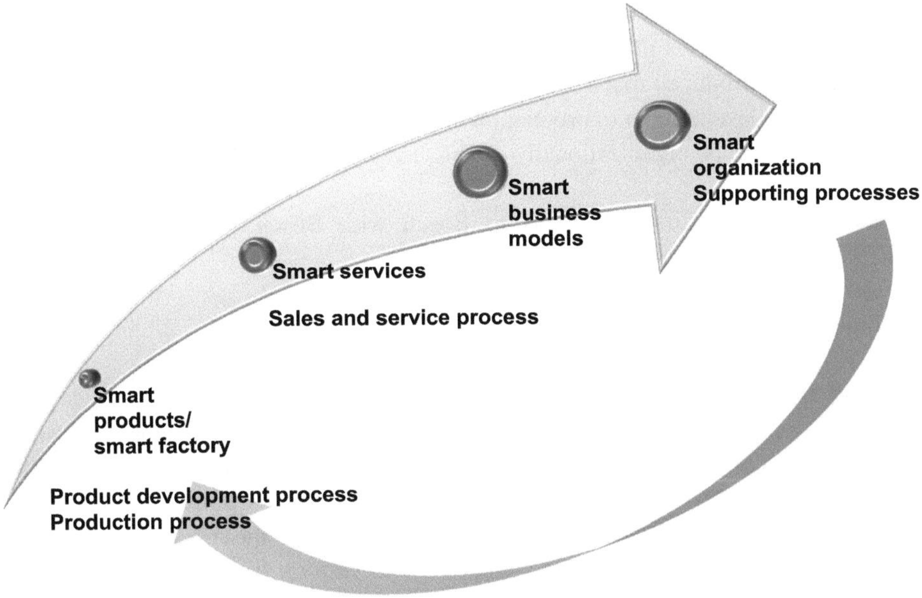

Fig. 9.3 Digital transformation implemented step by step

more and more companies. Digitalization is one driver here, while another is the steadily increasing regulatory requirements.

A schematic representation and classification of the individual transformation steps are shown in Fig. 9.3. In practice, there will typically be a much greater overlap of the individual phases. Without smart products, there can be no smart services. Often, the process also starts in isolation for existing products in production. Figure 9.3 suggests a sequential process. It should be understood that it is more of a cyclical endeavor, as indicated by the second arrow. This results from multiple products with different smart characteristics or stages of development. The customer benefit, which has already been mentioned several times, in combination with a cost-benefit analysis, determines the priority and thus the direction.

The topic of a paperless company is realized indirectly through the individual stages. A smart factory, for example, should be largely paperless. The same applies to the other stages and topics. Thus, at the end of the transformation, the company will have a very low paper consumption. Reducing it to zero is likely to be difficult, if only because there will always be partners, suppliers, or customers who, at the end of the day, still want to see paper.

In product development, approaches such as design thinking and agile methods should be used (but not only there). The strong interaction with other areas, such as production, sales, and service, which has already been mentioned several times, should actually have been implemented for years. Starting points for change are new products that are developed with new software solutions and agile processes. This should not necessarily be the most complex undertaking in the company. Especially with new tools, new processes can be introduced much more easily. Tool- and process-based vertical integration from product development to production is therefore essential.

The new areas can be organizationally established as

- an independent division, analogous to Bosch with Bosch Software Innovations, including profit and loss responsibility,
- a center of excellence without profit and loss responsibility,
- the founding of or cooperation with start-ups,

see [22]. All approaches have advantages and disadvantages. The "right" choice depends very much on the corporate culture. Alternatively, the founding of start-ups can also be considered, as practiced by Allianz, Zeppelin, or Oetker, for example, see [33, 34]. This can significantly increase the speed of change and avoid restrictions imposed by the company's own culture. However, one should not fall into the misconception that these "speedboats" can and will change the company's own culture. In order to be successful as a start-up, it must at least be operationally decoupled from the existing business. However, it is also necessary to integrate the innovations achieved in start-ups into the company itself—a challenging task. In doing so, reservations that exist on both sides often have to be overcome.

Regardless of the chosen organizational form, there will certainly be a coexistence of traditional and new/smart products, including their business models, for several years, as the transformation simply takes time. This does not necessarily make things easier. This affects both people and the company structure.

9.2 Transformation of Leadership Behavior and Managers

Digital transformation is a matter for top management! Unfortunately, managers at the "C-level" with a pronounced understanding of digitalization are still rarely found in Europe. Only 1% of non-executive directors on the supervisory boards of the largest European companies are recognized digitalization experts, see [3]. The situation is no better in leading DAX companies. Here, only a very few supervisory board members are computer scientists or possess the relevant skills, see [23]. Unfortunately, this lack of digital skills has not significantly improved in recent years.

9.2.1 Characteristics of Digital Leaders

Managers who have made digital transformation their mission (i.e., digital leaders) must possess different management skills and personality traits than their "analog" colleagues (i.e., "traditional" board members). They must have courage, assertiveness, intellect, and communication skills. At the same time, a high degree of innovative capability is required, meaning the ability to think "out of the box." They must always be ready to abandon existing approaches. Disruptive strategies must be pursued energetically and consistently. At the same time, strong communication and persuasion skills and interpersonal understanding are needed to win over opponents (of which there will be many) or to bring opinion leaders to their side. Due to the increasing internationalization of companies, these managers must develop more and more intercultural skills. The upcoming changes therefore require a very high level of leadership qualities. They are the driving force behind developments and, at the same time, must also act as mentors for their team members. In addition, a strong drive to achieve is essential. All in all, this is an extremely challenging management profile. However, this group of people or individual also has one of the most interesting and, above all, the most important and demanding task to accomplish within the company, see also [3]. One must not be discouraged by internal resistance. Such resistance will arise from a variety of areas.

9.2.2 Leadership Behavior

Changes are required at all management levels. Leaders must increasingly master and manage:

- Complexity,
- Flexibility,
- Innovation,
- Change

The complexity arises primarily from the greatly increased connectivity, growing globalization, increasingly dynamic markets, and the significant rise in information processing, including its methods and tools. Knowledge of how to handle data and assess new technologies and their possibilities is of central importance here.

The pace of technological innovation is steadily increasing. As a result, leaders must be able to adapt flexibly. Agility plays an important role in this context. Here, agility means the ability to recognize changes early and respond quickly, see [28]. It also means breaking away from previous thought patterns and exploring new paths.

Software development responded some time ago to the constant changes in companies. Concept phases lasting several months are now hardly conceivable or feasible and should therefore be a thing of the past. The same approach should be taken in other areas. As innovation becomes increasingly important, leaders must create the right conditions for their employees to be creative and innovative. This is an approach already known from the Lean world, see [12]. Regardless, existing knowledge must be questioned and, once again, cross-functional collaboration is required. This cross-disciplinary dialogue is becoming ever more important. Above all, this involves knowledge transfer. This also includes increasing the willingness to experiment and occasionally tolerating failures.

Primarily, changes in the workplace should come from within the ranks of employees. Change can only succeed with the involvement and participation of employees and the works council. Employees generally do not want to be changed, but rather to help shape change. This is an experience from transformation projects and is therefore independent of digital transformation or digitalization. Managers must provide the framework for this so that employees can act accordingly. Currently, such activities are referred to as "democratization of the workplace"; in the 1980s, such activities were called KAIZEN or continuous improvement. But even then, the term democratic workplace existed, see [12]. By involving employees from the outset, mistrust is reduced and, as experience shows, the acceptance of the works council increases significantly. This is the path taken by Bosch, for example. They asked their own employees how their business models could be attacked. Around 1,800 ideas were collected from the workforce and the best were implemented accordingly, see [20].

As already mentioned, the stability of companies is decreasing. This results, among other things, from changing business models and products. Thus, the pressure for change is increasing. As already described in Sect. 9.1.3, although the HR department should actively manage change, the primary responsibility of leaders is to prepare their teams for the future and bring them along on the journey. The new technologies require completely different qualifications from employees. The transition to the new will certainly not be smooth. Therefore, a certain degree of perseverance is also required. On the other hand, it is important not to overwhelm one's own employees. As a rule of thumb, new things should not account for more than 40%, otherwise resistance will arise.

Existing practices of internal decision-making must also be questioned. Often, especially strategy discussions and decisions are made within a narrow "inner" circle. Deliberately opening up and avoiding redundant knowledge leads to significantly better results. The same applies to group decision-making processes. Here, factors such as deference to authority, conformity, group pressure, and the destructive handling of bad news hinder work, see [18].

Furthermore, management itself must model the change. Only in this way can employees be encouraged to emulate, otherwise a credibility gap arises. Change becomes most tangible for employees through concrete goals. For example, in 2009, the CEO of Axel Springer communicated that in 10 years, half of the company's revenue and profit should come from digital business. In 2015, 62% of revenue and around 70% of EBIT came from digital activities, see [20].

9.2.3 Decision-Making

It becomes clear that digital business transformation is, quite simply, a management revolution. Managers and leaders must therefore fundamentally change themselves and their daily working methods. New technologies such as Big Data/Data Lake or Artificial Intelligence make data-driven decisions possible, and indeed, absolutely necessary. Unfortunately, this approach is countered by the phenomenon of "algorithm avoidance," that is, the avoidance or rejection of algorithms and their decisions. This phenomenon has been verified and proven in many studies. People tend to prefer human judgment—their own or that of others (such as consultants or colleagues)—over algorithms, and thus often make the wrong decision, see [8]. This, of course, also applies to leaders. Thus, a central task for leaders is to get themselves and their employees to trust data and algorithms (as already mentioned). There are several strategies for this, see [8]. Unfortunately, it is not enough to simply communicate the advantages of new algorithms to employees on a rational level. Such systems usually deliver better results than humans would. Unfortunately, however, machines also occasionally make mistakes. These are then often used as a reason for further rejection. The result is, quite simply, a loss of trust. Interestingly, human errors do not lead to a loss of trust to the same extent. This acceptance gap described above can certainly be reduced by using large language models such as ChatGPT. Such models also help in the transformation process, not just in specific applications. This makes the use and benefits of AI applications more tangible for employees.

Another approach to increasing system acceptance is to make them more human-like. Robots or speech recognition systems are given names. Siri from Apple is probably the best-known example. The aforementioned language is another attribute for humanizing systems. Here, the human tendency toward anthropomorphism—that is, the inclination to attribute human characteristics to things—is utilized, see [8]. Examples can be found in films, such as Star Wars with C-3PO and R2-D2 or DATA from Star Trek.

9.3 Employee Transformation

Digitalization has already transformed our entire working world today. Computers, tablets, and smartphones have become integral parts of our daily work routine. This applies to both direct areas (such as production) and indirect areas (such as traditional administrative departments), albeit to varying degrees. Employees must prepare themselves for lifelong learning, not only in academic fields. In times of digital transformation, this is becoming increasingly important even in direct areas. Unfortunately, this realization is often lacking, which can have fatal consequences for those affected.

It is therefore necessary to create appropriate learning opportunities for all areas and hierarchy levels, along with sufficient time to take advantage of them. However, employees must also be willing to embrace new trends and expand their own knowledge base accordingly. In practice, this is precisely where one of the main problems lies. Unfortunately, this statement applies to both employees and employers. It makes little sense for companies to provide learning opportunities if no time is allocated for employees to use them. Allocating dedicated learning time could help here. Through suitable learning platforms, the uptake of learning opportunities and learning progress can be monitored.

Regardless of these framework conditions, companies must also make use of this newly acquired knowledge. A prerequisite for this would be the corresponding know-how at the management level. Unfortunately, this is still very inadequate today. This also applies to supervisory boards. The topic of learning and employee qualification deserves high attention. Without successful further training and qualification throughout the company, digital transformation will fail. Therefore, it is necessary to create innovative learning opportunities. Gamification and new learning formats (e.g., learning nuggets of 5–10 minutes in length, i.e., “learning to go,” or podcasts) need to be developed.

9.3.1 Indirect Areas

In the future, performance should count, not presence. While this demand is obvious for indirect areas, such approaches are not applicable to direct areas (such as production). Care must be taken to ensure that the “gap of perceived unequal treatment” between the two areas does not widen too much. In addition, works councils often slow down overly progressive approaches. Nevertheless, companies must strive to be attractive to new employees. However, this is not achieved simply by setting up foosball tables or similar amenities. Modern workplaces and the increasingly demanded compatibility of work and family are requirements voiced by employees in all areas. Therefore, there is no way around more flexible models. This requires appropriate mobility offerings for employees. Employees are increasingly present in the office only a few days a week or month. The rest of the time, they spend on the road or working from home. The fact that such approaches work was demonstrated to us by the coronavirus pandemic. As already mentioned several times, more and more employees are demanding this kind of flexibility. This requires the consistent use and training in remote work tools, see [30]. These new work models require ever-increasing networking, which is supported by the aforementioned collaboration solutions. However, the current trend runs counter to this demand. The proportion of employees practicing full or part-time home office is continuously decreasing from 24 to 22%. In Germany, the share lags behind the global trend at just 11%, see [13]. This statement applied to the year 2017. In 2023, two-thirds of employees in German companies wish for a hybrid approach between home office and office work, and 10% can even imagine working entirely from home. Unfortunately, there is a marked distrust of home office in management circles. As a result, even after the coronavirus pandemic, 39% of employees still have to go to the office every day, see [27].

In the future, workplaces should offer employees and teams the opportunity for individualized and effortless collaboration. The personalization of digital technologies is essential so that employees can self-organize within teams, see [21]. Cloud-based approaches are certainly a good option here.

Due to increasing digitalization, a significant shift in jobs is expected. 30–50% of employees in Germany are engaged in more or less routine tasks. These are highly threatened by the advance of intelligent systems. Companies are therefore called upon to support their employees on the path toward digitalization and to provide appropriate further training. Upskilling their own employees strengthens a company's competitiveness and ultimately secures employment, see [32]. Specifically, such solutions offer productivity potential of 10–20%, see [17]. The greatest potential lies with so-called knowledge workers such as engineers or lawyers. Here, appropriate software can assist, for example, in contract reviews, see [2].

9.3.2 Direct Areas

People must be deployed according to their strengths, which lie in the rapid comprehension of unforeseen events. This applies especially to direct areas. Modern shop floor management increasingly consists of creative employees.

Simply increasing the degree of automation may superficially lead to economic advantages, but at the same time, it has also resulted in a decrease in flexibility, which in turn has affected profitability. With novel AI-based sensitive robots, there is a significant increase in flexibility, but humans must increasingly take on control tasks. The shift in tasks already familiar from the lean transformation at the beginning of the chapter thus becomes concrete here. Production processes are simplified for employees, monotonous activities are eliminated in favor of much more varied but also more demanding tasks, and above all, health burdens will continue to decrease, thus enabling older people to remain productive until reaching retirement age. As a consequence, there is a clear need for further development of employees. The scope of "simple" tasks will continue to decline. As a result, job profiles will change significantly in the coming years.

The increased need for control by employees also results from the fact that machines or systems can provide contradictory information. Here, people are needed with their professional expertise as well as their general problem-solving skills to correctly assess the situation and, based on that, make the right decisions.

To meet the significantly increasing demands, employees must be supported in their production tasks by assistance systems. The share of labor costs and thus the pressure on wages will decrease, as China and other low-wage countries will lose significance. This results from the rising wage levels in these countries, but on the other hand, a lot size of one from China simply is not worthwhile. This should also somewhat reassure the unions and hopefully contribute to greater acceptance of smart technologies.

Another element, which has hardly been discussed so far, is the "digitalization" of the human being, i.e., the human becomes a CPS themselves. For this purpose, each employee receives a digital representation (not in the form of an avatar, but simply as a file containing production-relevant information). The information could include, for example, body height, physical restrictions, changed living conditions, individual preferences, or employee skills. With the help of this data, production systems can assign employees better and more optimally. Just knowing the body height allows assistance systems to be automatically set optimally. Various assembly alternatives (sitting or standing) can also be adopted, including the time required (which can certainly differ between the two variants). These possibilities go far beyond previous approaches, see [14].

By using predictive activities (e.g., predictive maintenance), work for maintenance becomes much more plannable, the need for on-call duty will be significantly reduced, and travel activities (especially short-term) will decrease. In this context, the correct representation of location information (i.e., where exactly a person, machine, or product is located) is becoming increasingly important. This leads to further optimization of maintenance work. In general, data-driven approaches, especially in maintenance, offer significant economic potential, such as reducing spare parts inventory.

Current research is also attempting to virtualize activities in production, see [25]. These are initial steps in this direction. Such approaches are currently being funded by the German Ministry of Economic Affairs. In the coming years, initial research results can certainly be expected. However, flexibility can already be expanded today through dynamic shifts. Employees could independently assign themselves to flexible shifts using a smartphone (bring your own device). In addition, special shifts (for example, on a Saturday) could be "advertised," for which employees can then apply.

Overall, each individual employee must develop their own "strategy" for dealing with machines and increasingly complex IT systems, with the goal of not being "replaced" by a machine or IT system. This can range from specialization to the already mentioned cooperation, see [6]. But regardless of personal and company strategy, the topic of digitalization and the increased use of robots in production as well as in administrative and management areas cannot be stopped. Thus, adaptation and differentiation are required for everyone, regardless of industry or career level.

9.3.3 Approaches for Action

Further training should not be limited to the academically educated employee sector. In particular, skilled workers, who form the backbone of every production, must be appropriately supported, including the identification of specific career paths. Appropriate in-house and external offerings need to be developed here. This is also necessary because the content of existing job profiles is changing, but new ones are also emerging, and consequently, existing job profiles are disappearing. Career paths will also change. A few years ago, data scientists were absolute rarities, but currently, almost all automobile manufacturers are looking for employees with such qualifications (regardless of what exactly is meant by this term).

The changes in companies will be very profound. Therefore, good cooperation with unions and works councils is essential. Only together with the workforce can the future be built and shaped positively. However, excessive regulation within companies should be avoided. Flexibility is key.

References

1. Markus Albers, Der Chef im Glaskasten, Lufthansa exclusive No 3, 2016.
2. Klaus Becker, Preise für visionäre Firmengründungen, TUMcampus, 3/2017, Seite 36–37.
3. Thomas Becker, Carsten Knopp, Digitales Neuland: Warum Deutschlands Manager jetzt Revolutionäre werden, Springer Gabler, Oktober 2015.
4. Martin Böker, Mobiles Arbeiten: Enterprise Mobility: Ran an die Fachbereiche, Computerwoche, 25.07.2017.
5. Michael Chui, James Manyika, Jacques Bughin, Richard Dobbs, Charles Roxburgh, Hugo Samazin, Geoffrey Sands, Magdalena Westergren, The social economy: Unlocking value and productivity through sozial technologies, McKinsey Global Institute, Juli 2012.
6. Thomas H. Davenport, Julia Kirby, Dein Freund der Roboter, Harvard Business Manager, 9/2015, Seite 22–31.
7. Jens Fath, Erfolgreiche Transformation ist Business-getrieben, ManagerWissen, 04/2023.
8. Walter Frick, Wenn der Chef Hosen aus Metall trägt, Harvard Business Manager, September 2015, Seite 42–49.
9. Gary Hamel, Worauf es jetzt ankommt, Wiley-VCH Verlag, 2013.
10. Sven Michael Hoffmann, PwC-Studie: Die Rolle des Chief Digital Officers ist auf dem Vormarsch, 14. Dezember 2022.
11. Laura Hopp, Lynn-Kristin Thorenz, Matthias Zacher, Unternehmen zwischen Tradition und Wandel: Die digitale Transformation in Deutschland, IDC White Paper, April 2017.
12. Masaaki Imai, KAIZEN – Der Schlüssel zum Erfolg im Wettbewerb, Econ-Verlag, 1986.
13. Josef Joffe, Zeitgeist: Ab ins Büro!, Die Zeit, Nr. 32, 3. August 2017, Seite 10.
14. Christoph Krieger, Holger Müller, Der Mensch als cyber-physikalisches System, Beschaffung-aktuell, 7–8/2015.
15. Ralf T. Kreutzer, Treiber und Hintergründe der digitalen Transformation in D. Schallmo (Hrsg.), Digitale Transformation von Geschäftsmodellen, Springer-Verlag, 2017, Seite 33–58.
16. Gina Luk, Strategy Analytics: The Global Mobile Workforce is Set to Increase to 1.87 Billion People in 2022, Acounting for 42,5% of the Global Workforce, 9. 11. 2016, http://www.prnewswire.com/news-releases/strategy-analytics-the-global-mobile-workforce-is-set-to-increase-to-187-billion-people-in-2022-accounting-for-425-of-the-global-workforce-300359-376.html, aufgerufen am 27.07.2017.
17. J. Manyika, M. Chui, P. Bisson, The Internet of Things: Mapping the value beyond the hype, McKinsey Global Institute, https://www.mckinsey.de/files/unlocking_the_potential_of_the_internet_of_things_full_report.pdf, aufgerufen am 23.08.2017.
18. Kurt Matzler, Frank Bailom, Stefphan Friedrich von den Eichen, Markus Anschober, Digitale Disruption: Wie Sie Ihr Unternehmen auf das Digitale Zeitalter vorbereiten, Verlag Fran Vahlen, 2016.
19. Andrew McAfee, Erik Brynjolfsson, Die große Abkopplung, Harvard Business Manager, September 2015, Seite 32–41.
20. Jürgen Meffert, Heribert Meffert, Eins oder null, Econ Verlag, April 2017.
21. Thomas Muhr, Kommunikation und Collaboration – Der Arbeitsplatz der Zukunft setzt neue Technologien voraus, Computerwoche, 26.07.2017.
22. Michael E. Porter, James E. Heppelmann, Wie Smarte Produkte den Wettbewerb verändern, in Harvard Business Manager, Dezember 2015, Seite 52–73.
23. Dietmar Palan, Gesucht: bunte Hunde, manager magazin, August 2017, Seite 23.
24. Alec Ross, Die Wirtschaftswelt der Zukunft, Plassen Verlag, 2016.
25. Peter Sauer, Industriebarbeit vom Küchentisch aus, Tagesschau, 28.07.2021.

26. Frank Schabel, Kultur zum Frühstück, Harvard Business Manager, 4/2017, Seite 28–29.
27. Felix Seifert, Dan Honigstein, Jeder zehnt wünscht sich hierzulande, gar nicht mehr ins Büro zu kommen, Welt, 23.06.2023.
28. Günther Schuh, Wie agil ist ihr Unternehmen, Harvard Business Manager, 4/2017, Seite 62–63.
29. Esther Shein, Woran die digitale Transformation scheitert, CIO, 08.12.2022.
30. Charlotte Trueman, Mitarbeiter wollen Remote-Work-Tools, keine Office-Pflicht, CIO, 27.06.2023.
31. Frank Weber, Joachim Berendt, Robuste Unternehmen: Krisenfest in Zeiten des Umbruchs (essentials), Springer Gabler, Mai 2017.
32. Horst Wildemann, Produktivität durch Industrie 4.0, 24. Münchner Management Kolloquium, 14./15. März 2017.
33. Horst Wildemann, Neue Geschäftsfelder. Geschäftsmodelle, Technologien: Müssen sich Unternehmen neu definieren? TCW-report Nr. 69, Verlag TCW Transfer-Centrum, 2017.
34. Oetker investiert in Start-Up, Lebensmittelzeitung, 01.09.2017, Seite 17.
35. ReportLinker, Global Mobile Workforce Solutions Industry, 26. Januar 2021.
36. Bundesministerium für Wirtschaft und Klimaschutz, Den digitalen Wandel gestalten, 09. Dezember 2022.

10 Risks From New Competition

Examples have already been given of established companies being displaced by new market players. Furthermore, the statement holds true that everything that can be digitized will be digitized. Thus, all industries are affected by digitalization and are therefore threatened by new competition. Risks must therefore be identified at an early stage and appropriate countermeasures initiated. In addition, a continuous competitive analysis must be conducted.

Up to now, industrial revolutions in Germany have been the foundation of its prosperity. Only if Germany masters the development towards digitalization and Industry 4.0 can its strong industrial culture be preserved in the long term. Currently, Germany is at risk of falling behind. At the national level, there is a "Don't change a running system" attitude—things are going well for us, so why should we change anything? This attitude is extremely dangerous given the profound changes taking place and threatens the prosperity of us all. Above all, Germany continues to lag behind its potential, especially in the public sector.

Specific disruptive changes threaten established companies at their very core. This threat results from their characteristics (as already mentioned), which are initially not very appealing:

- unclear prospects and market potential,
- starting points usually in niche areas,
- after an initial phase, rapid and sometimes unchecked growth,

See [10]. The low attractiveness in terms of market volume threatens established and large companies in the market. Large companies require correspondingly large market potentials. This resulted in the initial reluctance of German car manufacturers regarding electromobility. For a long time, the technology fit exactly the characteristics described

W. Huber, *Industry 4.0 Compact—How Technologies Are Transforming Our Economy and Our Companies*, https://doi.org/10.1007/978-3-658-51029-9_10

above. On the other hand, it is precisely the characteristics of disruptive changes that show that investments should be made in these segments. BMW is taking this path with its “i” series. The situation has now changed significantly. Both the car manufacturers have recognized the trend towards electromobility, and regulatory frameworks have also contributed to this development.

The situation described makes clear and at the same time justifies the assessment of John Chambers (Chairman of CISCO) that around 40% of today’s top companies will not survive the next 10 years, see [10]. Dr. Gerhard Kebbel (former Head of Digitalization at the Hessian State Bank and current board member at Frankfurter Bankgesellschaft) shares a similar view: “If you can foresee a disruption, then it is no longer a disruption,” see [5].

However, non-disruptive changes should not be overlooked. In this case, companies open up entirely new business fields and markets. These are outside existing industries, see [7]. Thus, companies must focus on both types of innovation, which does not necessarily make things easier. Microloans, windshield wipers, dishwashers, or Viagra are just a few examples of such new products and thus new business fields, see [7]. Experience in recent years has shown that it is precisely these non-disruptive innovations and changes that have a significantly greater impact than disruptive changes.

10.1 Examples of New Competition

All industries and thus all companies face dangerous competition from new players. The financial sector is one example. While startups in this sector—fintech companies—were only able to raise US$930 million in financial support worldwide in 2008, the amount rose to US$3 billion in 2013 and to around US$8 billion in 2018, see [15]. In 2022, the market value of the 10 largest companies in this sector exceeded US$250 billion, see [12]. This shows that more and more investors believe in their business models and also highlights the transformation the industry has been undergoing for years.

The ride-sharing service Uber had an estimated value of around US$69 billion in 2016, and by 2023, despite some turbulence, it had already reached US$88 billion, making it worth twice as much as the two established car rental companies Hertz and Avis combined. Investors here include, among others, Google Venture, see [15], [17]. At the same time, in 2016 the company posted a significant loss of US$2.8 billion on sales of around US$6.5 billion and had burned through about US$8 billion since its founding, see [19], [17]. The company also failed to generate an operating profit in 2022.

The industry leader in online business—Amazon—is also developing into a hardware supplier, among other things. The company and its business model do not represent a disruptive approach. Even at Amazon, payments are still made with money. Rather, it is perfecting the art of selling. For Amazon, it is about building its own digital ecosystem. Amazon manufactures its own eReader with the Kindle. In addition, the company is also the leading provider of cloud solutions, produces its own movies and TV shows, is entering the grocery sector, and offers Alexa, a voice-controlled solution for the smart home. The company also works very intensively with big data solutions. Amazon generates about one third of its revenue through personalized recommendations, see [8]. The goal is to extend its own value chain as far as possible.

Another example is the music industry. Since its peak in 1999, it lost around 70% of its revenue by 2014, with the trend continuing downward, see [10]. This is the result of various new products such as Apple's iTunes and streaming services. Customers no longer have to buy entire albums, but only the tracks they actually like. Even this is becoming obsolete with streaming services. Monthly flat rates allow unlimited consumption of music for a fixed price; usage is now the focus, not ownership.

10.2 Causes of New Competitors

The causes of new competition are multifaceted, see Fig. 10.1, which illustrates the various possibilities.

Digital development is often exponential and combinatorial. Existing or new technologies are combined to create new solutions and new business models, as the examples in the individual chapters have shown, among others. The threat can therefore come "from all sides." As a result, specific industry knowledge is subject to significant inflation, and industry boundaries are increasingly disappearing. Consequently, entry barriers are reduced. However, this is not an automatic process. As a consequence of digitalization, there is also a democratization of the means of production, see [8]. This movement

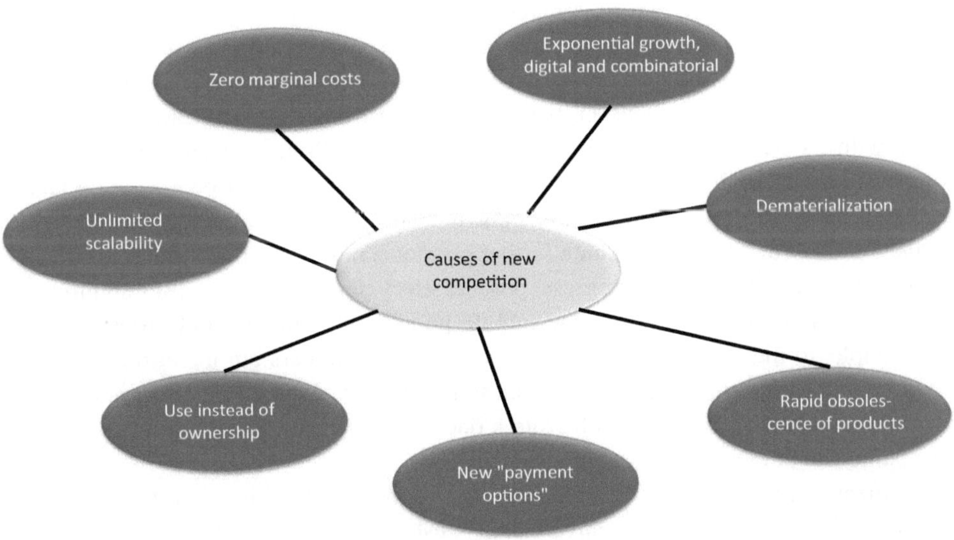

Fig. 10.1 Causes of new competition

is led by 3D printing, i.e., additive manufacturing. This trend has also been evident in the music industry for several years. Self-produced music distributed via YouTube has already been the starting point for a successful career for some artists. At the same time, it has become standard for many artists to be present on social media. It is evident that traditional music companies are increasingly receding into the background, and artists are moving towards greater self-marketing and thus emancipation. This is also reflected in revenue trends, see Sect. 10.1.

The increasing digitalization is bringing about a paradigm shift. The "old" economy was and is characterized by economies of scale. The more units of a product are produced, the lower the price becomes. In contrast, the digital economy is characterized by network effects and combination, see Chap. 8. The success of digital business models, as has been mentioned several times, results, among other things, from zero marginal costs, but also from the extremely high speed and reach enabled by the dematerialization of products (this does not apply to all digital business models, of course).

US companies such as Twitter, Facebook, WhatsApp, and Amazon demonstrate this approach impressively, especially in the B2C environment. It also becomes clear that expansion does not necessarily have to be capital-intensive. Pokemon Go has shown this. Three weeks after its release, it had already been downloaded over 75 million times, see [10]. This sometimes enormous speed through network effects is also responsible for the "The Winner Takes It All" phenomenon. Competitors no longer have time to respond accordingly. It is also evident that such disruptive new business models are primarily generated by start-ups and not by established companies. On the other hand, around 90% of German companies consider innovations and new business models to be important to very important. However, these are primarily product innovations. Only one in ten companies is working on new business models or services, see [22]. A lack of cooperation between companies is certainly one of the causes here.

Another reason for the enormous speed is the aforementioned dematerialization and the shift from ownership to usage-based approaches. Uber demonstrates this impressively with its business idea. The company is the world's largest passenger transport company, present in 70 countries and over 400 cities, but does not own a single taxi or similar vehicle, see [10]. Dematerialization is also associated with high scalability of business models and products, which further promotes rapid dissemination.

Due to rapid technological progress, product lifecycles are in some cases significantly reduced. This applies not only to the mobile phone and computer industries, but also, for example, to the automotive industry. As a result, customers are increasingly tending to use functionality exclusively rather than to purchase it. A nice side effect is the lower financial burden. Expenditures shift from a one-time expense to ongoing costs, which is interesting and attractive for many consumers.

Payment methods are also changing. Instead of money, there is an exchange of product or product use for data or advertising. Customers often accept advertising if, in return, they can apparently use online services for free, or they provide user or location data to the product provider.

Another important point is the enormous growth of American technology companies, especially the "Big Five" (Apple, Amazon, Google, Facebook, and Microsoft). These companies have been able to develop an overwhelming market dominance in some areas. For example, Google practically dominates internet search and is very strong in the field of smartphone operating systems. In the cloud sector, Amazon dominates the market with around 44%, Microsoft with 11%, and Google with 6% market share. With very high profit margins of 16% at Amazon and up to 70% at Apple, these companies also have enough liquidity to try out new business ideas as well as to acquire interesting companies, see [2]. Such behavior is even expected of them. This is also reflected in their enormous growth, for example, 23% at Amazon and 17% at Google in 2017, see [2]. The current geopolitical changes have also left their mark on the tech giants. Growth rates are melting away like snow in the sun, and layoffs are no longer a taboo subject.

Traditional companies find it difficult to compete with this. The unbroken market power is also reflected in their behavior towards suppliers and customers.

However, the example of Netflix also shows that a misjudgment of the market can lead to a fatal misdevelopment. The company wanted to monopolize television and failed spectacularly. The company lost around 60% of its market value, see [14].

Hardly any company can resist the "Big Five" anymore, as the example of Nike (as market leader in the sports goods sector) in its relationship with Amazon shows. Furthermore, due to their business models, these companies have such extensive knowledge about their customers that they could successfully enter virtually any business field, whether it be insurance or autonomous driving. Knowledge about customers thus provides the basis for new business ideas. The companies continuously renew themselves through this.

It is precisely this market power of the tech giants that is increasingly unsettling politicians. Thus, the current efforts of the US Congress and also of Europe to limit their strong dominance through regulatory measures are not surprising, see [9]. In Germany, too, calls for corresponding restrictions are becoming increasingly louder, see [11].

10.3 Identification of New Competitors

As has been mentioned several times, new competitors no longer emerge solely from the familiar environment. As a result, recognizing competition is becoming increasingly difficult. Disruptive approaches are characterized, among other things, by the fact that when they first appear, they seem inferior to established products and business models or address only a niche market. Consequently, they quickly disappear from the "radar" of established companies. However, when these new approaches succeed, they unfold their impact at an extremely rapid pace—see also the introduction to this chapter.

Competition threatens from:

- Start-ups,
- Market peers (also called competitors),

- Partners,
- Suppliers,
- Other industries,
- Other countries.

Companies must therefore continuously review the competitive landscape. Porter's competitive analysis is a good approach here, see [13]. On the other hand, as already mentioned, it is important to further develop one's own innovative capabilities in order to secure economic growth through non-disruptive changes.

At the beginning of a competitive analysis, understanding new technologies should be the focus. These often serve as the starting point for start-ups and potential disruptive approaches. Understanding and thus mastering new possibilities should become a core competence within the company. Only those who adapt quickly and embrace new ideas and opportunities will survive. By understanding technologies, an early warning system can be established and expanded, which represents a proactive approach. This also involves identifying one's own core competencies. These must be protected, appropriately positioned in the market, and used to generate a forward-looking and viable digital business model. This also includes involving one's own employees, partners, suppliers, customers, or external consultants. Through collaboration platforms, employees can contribute to the company's strategy. Over 70% of companies using such solutions employ them for market scanning, see [10]. Open Strategy or Open Innovation, as further approaches, can generate surprising and new ideas. However, for this to happen, a company must "step out of its comfort zone" and share information with the online community. Another option is hackathons. Such events can generate new ideas and solutions for existing issues. It is often said that the best employees do not work within one's own company. At least partially, this resource can be accessed in this way. Companies must therefore break out of traditional behavioral patterns if they want to create the "next big thing" or avoid being overtaken by it. It can be observed that the majority of business model innovations are not truly new, but rather transfers from other industries. Thus, cooperation with companies from other sectors is meaningful and advantageous for all parties involved. Above all, it protects against unexpected competitors.

The "playing field" for participants is overall becoming larger, as new rules apply and industry knowledge is not always required. Investors are ready to finance start-ups (at least in the USA and to a lesser extent in Germany). In addition, new "game pieces" are constantly being added to the board in the form of new technologies or their combinations. Furthermore, the internet is constantly bringing new "players" onto the board. Regional boundaries are increasingly disappearing. The entire environment is therefore highly dynamic and, at the same time, opaque. This applies to both disruptive and non-disruptive changes, as has been mentioned several times.

Participation in relevant venture capital companies provides access to new ideas and potential new competitors. Porsche and BMW, for example, follow this approach, also to bring new know-how into the company. Through cooperation with universities, which is often offered, one can gain insight into the start-up scene. This can also bring fresh knowledge into the company, and in some cases, even lead to the founding of an in-house start-up, as practiced by Airbus, for example.

10.4 New National Competitors

New competition is not limited to companies. The German business location as a whole is also facing challenges at the national level. The challengers include well-known countries such as the USA and now also China, but also countries like France or Switzerland. The following section focuses on these four countries as examples.

10.4.1 China

China is not only the largest sales market for automobiles, but also the largest production location. It is now also one of the largest markets for industrial robots. Consequently, the robot manufacturer Kuka was acquired by a Chinese company.

With the “Made in China 2025” strategy, the country aims to overtake the Western industrialized nations, especially Germany. The goal is to move out of the low-wage segment and produce top quality at affordable prices. For China, Germany is seen as the great role model to surpass. Many analysts see China as being on a good path in this regard. On closer inspection, however, it becomes apparent that many of the projects labeled as Industry 4.0 are, in reality, at best lean transformation projects or involve the introduction of SAP as an ERP solution, MES, or PLM systems. Making comprehensive use of these solutions is another matter entirely, and unfortunately, this often does not happen. The country also partially shields itself from foreign products. As a provider of industrial products, the market is therefore extremely challenging. In addition, the trade conflict with the USA is weighing on economic development. Moreover, it is increasingly apparent that subsidies alone are no guarantee for international cutting-edge technologies. On top of that, the approach is extremely costly. However, the coronavirus crisis has sharpened the focus on digitalization. In certain technologies, such as AI, the focus is strongly on specific solutions, such as surveillance software and thus image recognition. While China is catching up with Western countries in these areas overall, there is a lack of broad implementation.

However, in the field of software development and social media, Chinese companies are increasingly becoming competitors to their Western counterparts. In this area, the Middle Kingdom has already overtaken the USA. Companies like WeChat and Alibaba are now active worldwide. For example, in recent years Alibaba has invested around US$15 billion in expanding its global logistics network, see [20]. A social media revolution is underway in China, enabling a level of communication intensity previously impossible. The company Qzone already has 625 million users, followed by WeChat with about 355 million. In China, around 1 billion people own a smartphone, see [8]. Alibaba, the Chinese counterpart to Amazon and PayPal, is active in over 200 countries and regions and processes over 2.85 million money transactions per minute via its payment service Alipay. That is significantly more than PayPal, see [15]. Alibaba is certainly an exception here, as it operates from and within the protected Chinese market. Nevertheless, these figures give a clear impression of the size and intensity of the local market. Most of these owner-managed companies pursue a very long-term strategy focused on growth and market share rather than profit. In many cases, there is also state support.

China is also increasingly establishing itself in the smartphone and electronics market. Huawei is one example. Overall, China is a country with considerable industrial potential. Due to the state-controlled economy, local companies have significant advantages in the domestic market. However, these companies are increasingly operating internationally, making them serious competitors. In general, global innovations in internet technologies are being strongly driven forward in China and other Asian countries. These countries rank well ahead of Western nations such as the USA, the UK, or Germany, see [8]. A similar picture is emerging in the automotive sector. More and more Chinese companies are entering the European and German markets. Conversely, German car manufacturers are losing market share in China.

It is therefore advisable not to lose sight of the Chinese market when it comes to innovation. It is also a good example of the market dynamics in the world's most populous country. China is catching up in all respects. For example, ICBC ranks as the world's most valuable banking brand, ahead of Wells Fargo Bank.

10.4.2 France

Since the election of President Macron, the topic of digitalization and Industry 4.0 has gained significantly in visibility. Government efforts have been considerably intensified. However, initiatives such as "La French Tech" were already launched during his predecessor's term in 2013.

Investments in start-ups fit this picture. In Europe, France ranks second behind the United Kingdom, see Fig. 10.2 and [16]. For the years 2021 and 2022, investments are developing very positively. For Germany, this means over €17 billion in 2021 and €9.8 billion in 2022, see [24]. In France, more than €6 billion were invested in 2022, see [25].

Country	Investments in billion USD, 2016
United Kingdom	3.6
France	3.0
Israel	3.0
Germany	2.2
Sweden	1.8

Fig. 10.2 Country-specific investments in start-ups

It is striking that Germany is not among the leaders. Even the much smaller Israel can show higher investments, and Sweden is just behind Germany. Israel is generally regarded as a high-tech forge. In line with this, Cisco invested around €400 million in 2017 in a new French venture capital provider, Partech Ventures. However, this picture is highly dynamic and should therefore not be overestimated. France in general, and Paris in particular, are currently among the most attractive markets for start-ups and digitalization. French centralism is helpful in this regard. A so-called "Tech Visa" is also offered for innovative individuals to facilitate controlled immigration from abroad. In addition, the French government has launched a €10 billion program for innovation. The French president also personally promotes his own country and presents himself in the media—even in English. Local companies are investing nine-figure euro amounts in their own start-up scene, see [16].

10.4.3 USA

The USA boasts Silicon Valley, a high-tech forge of the highest caliber. Some of the world's most valuable brands are also based in the USA. The top five are Google, Apple, Microsoft, Amazon, and Facebook, see [21]. The figures vary somewhat depending on the time examined and the source. What stands out here is the dominance of technology companies. Even in the manufacturing sector, German automobile manufacturers are facing an increasingly serious competitor in Tesla. The highly dynamic and at the same time innovative American market is certainly also a result of the high levels of available and invested venture capital compared to Europe. In 2014, $52 billion was invested as venture capital in the USA, compared to €10.6 billion in Europe. In 2015, the situation was not much better, with $74.2 billion in the US versus $13.4 billion in Europe, see [10]. In 2021, the peak was reached with over $345 billion. A year later, the value had already dropped again to around $240 billion, see [23]. This demonstrates the high level of dynamism and thus the limited significance of such figures. Surprisingly, in this context, the USA is "only" among the early adopters when it comes to internet innovations, see [8]. In particular, the Inflation Reduction Act introduced by the current US government grants companies tax breaks and subsidies amounting to $369 billion, see [3]. This clearly demonstrates the efforts to further strengthen the country's industrial potential and to achieve production relocations.

10.4.4 Switzerland

Switzerland, as an absolute high-wage country, thrives on innovation. It is therefore not surprising that the Swiss have also launched an Industry 4.0 initiative called "Industry 2025." This was initiated by four industry associations. Analogous to its European neighbors, Swiss companies focus very strongly on the smart factory—that is, the optimization of their own value chain and less on the development of new digital business models, see [18]. Thus, the adherence to familiar and traditional patterns of thinking is unmistakable. This is all the more surprising given that the country has produced groundbreaking innovations in recent decades (such as the Pascal programming language, the online calendar Doodle, the scanning tunneling microscope, or coffee capsules). It also likes to refer to itself as the "land of inventors." The country ranks at the top of the global innovation index, see [4]. However, the examples also show that these are primarily physical products.

However, there is movement in the start-up sector. For example, the eHealth provider Visionarity AG was able to position itself at the top in a Europe-wide comparison and was able to acquire venture capital a few years ago, see [6]. Unfortunately, the company has been in liquidation since 2019. Start-ups and SMEs are not sufficiently supported in Switzerland. Thus, Switzerland does not differ significantly from its neighbors in a European comparison.

10.5 Defensive Measures

"Disrupt or be disrupted," a quote attributed to Mark Zuckerberg, aptly expresses the situation, see [10]. Companies are better off cannibalizing themselves than letting someone else do it. The key is to constructively question one's own business model. In the case of evolutionary developments, this is generally not a major problem. The situation is different with revolutionary or disruptive approaches. The discovery of new markets is becoming increasingly important due to digitalization. It enables the promotion of one's own economic development.

Regardless of whether the approach is evolutionary or disruptive, companies need to recalibrate their internal monitoring radar. Approaches for identifying new potential competitors were provided in Sect. 10.3.

Defensive measures should not be limited to a single, isolated action, but should instead comprise a bundle of activities and measures. Only in this way can competitors be kept at bay.

10.5.1 Self-Cannibalization

This approach sounds like a contradiction. Companies and managers are supposed to go against customer needs and opt for low margins, lower sales, and high risk. This is precisely where the challenge lies. As already mentioned, disruptive business models are often initially characterized by exactly these attributes. For start-ups, however, they pose no problem, whereas for established companies and especially market leaders, they are almost insurmountable obstacles—which is quite understandable at first glance. Start-ups do not even intend to compete with established companies at the beginning. In many cases, this would exceed their financial and management resources. One example is Tesla. The company deliberately started in the premium segment, with vehicles costing over €70,000. Here, it was possible to demonstrate that electric cars work, and in this price segment, customers were also willing to put "a little more" on the table for an electric car. Once the company was established, the next step with the Model 3 was to target the mass market. Porsche's strategy is moving in a similar direction. The company wants to become a "German Tesla." In Porsche's luxury segment, customers are indeed willing to pay more for an electric car (just as with Tesla). BMW is also pursuing a disruptive approach with its i-models. Both BMW and Porsche are taking a very arduous path. Implementing change internally is a grueling slog. This is also the reason why many companies shy away from this path.

The simplest and most pragmatic way to act disruptively is to found a start-up oneself, as the Klöckner Group did with Klöckner.i, see [10]. Both approaches (a separate division/brand like BMW and Porsche, or a start-up) can lead to success. Instead of founding their own start-up, companies can also participate in or acquire such firms. However, one should not attempt to integrate a start-up into one's own corporate structure. This approach is bound to fail.

The examples show the different implementations or interpretations of the 4-zone model, see [1]. In this model, there is a clear separation of company divisions:

- in the performance area, the company earns money with its existing products,
- the productivity area ensures further development of existing products,
- the incubation area serves as the company's "playground." This is where new ideas and products are tried out. It also serves as an early warning system,
- in the transformation area, new business models are developed, based on promising ideas from the incubation area.

BMW and Porsche pursue a very stringent implementation of this approach. Klöckner, on the other hand, interprets "areas" as separate companies (start-ups).

10.5.2 Building Your Own Digital Ecosystem

Apple and Amazon have clearly demonstrated the usefulness of having their own digital ecosystem. This creates high entry barriers for competitors and, at the same time, high switching costs for their own customers. The high entry barriers especially deter start-ups. In this way, market power can be expanded. These entry barriers can be reinforced by broadening the product definition. By networking their own products, a product system or a system of systems can be established, see [13] and Chap. 6. This also includes using customer data for further innovations. Newcomers may have new business ideas, but they do not have access to the same wealth of data and thus the knowledge about customer needs. However, this should not lead to focusing exclusively on evolutionary improvements. The consistent use of (customer) data, not only for technical innovations but also for new business models, thus represents an important lever. This brings us back to non-disruptive approaches.

10.5.3 Leveraging Your Own Strengths

For large corporations or traditional medium-sized companies, it makes little sense to try to copy start-ups. Instead, companies should focus on their own strengths. These include, among other things, their usually strong market position and their own brand—that is, brand image. This is one of the strategies used by German car manufacturers. Market newcomers may have an innovative image, but at the end of the day, especially in the automotive industry, customers prefer to spend their money on the prestigious established manufacturers' expensive products. Customization of products is also part of the brand image. This creates an emotional bond with the company and the product. However, recent developments in the automotive industry show that one should not rest on these "laurels." The long-ridiculed Chinese manufacturers are becoming increasingly confident and are expanding beyond China, including into Germany.

Open standards can also be among these strengths. In the field of agricultural technology, for example, a closed digital ecosystem is hardly conceivable. The same applies to the smart home sector. Here, German companies can leverage their technological strength and ability to cooperate.

10.5.4 Organizational Measures

Regardless of the defensive measures listed so far, companies must become more robust. This means increasing their own innovation potential and adaptability, and viewing the rapid adaptability of their business models as a strategic competitive advantage. If this is lacking, these companies will fall behind, with the corresponding consequences. This requires questioning one's own business model. The "Nightmare Competitor" approach can help here, see [10]. In this approach, a worst-case scenario is developed to determine what maximum damage to one's own business model could look like. However, a company organization must be willing to adopt this approach and, subsequently, to draw the necessary consequences.

The fast mover, first mover approach goes in a similar direction. Here, the company establishes itself as a trendsetter in terms of technologies and business models. If this approach becomes part of the "corporate DNA," the company continuously reinvents itself. This is the path taken, at least in part, by the American "Big Five." A willingness to experiment, combined with the possibility of failure, initially incurs low costs but reduces the risk of being overtaken by disruptive competitors.

In addition, companies must work on their operational efficiency, see [13]. This means optimizing processes from product development through production to service and aftersales. However, the example of Uber also shows that management as a craft has not lost its value. Uber lost significant market value due to poor management, which resulted in a drop in the company's value, see Sect. 10.1. The car manufacturers demonstrate another alternative. They joined forces and jointly acquired the map provider Here from Nokia. Cooperation with established companies can strengthen one's own market position and, in turn, increase market entry barriers.

The individual approaches listed here are not mutually exclusive; on the contrary, they actually complement and reinforce each other. As has been mentioned several times, disruption is only one facet. Non-disruptive approaches can also enable very strong growth.

References

1. Ölmer Atiker, In einem Jahr Digital, Wiley-Vch Verlag, März 2017.
2. Mark Böschen, Trillion Dollar Babies, manager magazin, August 2017, Seite 88–94.
3. Gerald Braunberger, Die neue amerikanische Herausforderung, Frankfurter Allgemeine Zeitung, 07.12.2022.
4. Thomas Fuster, Die Schweiz erfindet sich ständig neu, Neue Züricher Zeitung, 15.08.2016, https://www.nzz.ch/wirtschaft/kommentare/innovation-im-vergleich-die-schweiz-erfindet-sich-staendig-neu-ld.111096, aufgerufen am 13.09.2017.
5. Laura Hopp, Lynn-Kristin Thorenz, Mattias Zacher, Unternehmen zwischen Tradition und Wandel – die digitale Transformation in Deuschland, IDC White Paper, April 2017.
6. M. Jordan, Visionarity als Top-Startup unter 500 in Europa ausgewählt, 5. Juli 2017, Land der Erfinder: Das Magazin für Innovationen, http://www.land-der-erfinder.ch, aufgerufen am 13.09.2017.
7. W. Chan Kim, Renee Mauborgne, Wie entsteht was neues, Harvard Business Manager, 07/2023.
8. Ralf T. Kreutzer, Karl-Heinz Land, Digitaler Darwinismus: Der stille Angriff auf Ihr Geschäftsmodell und Ihre Marke, Springer Gabler, Februar 2016.
9. Roland Lindner, Amerika knöpft sich seine Tech-Giganten vor, Frankfurter Allgemeine Zeitung, 12.06.2021.
10. Kurt Matzler, Frank Bailom, Stefphan Friedrich von den Eichen, Markus Anschober, Digitale Disruption: Wie Sie Ihr Unternehmen auf das Digitale Zeitalter vorbereiten, Verlag Fran Vahlen, 2016.
11. Friedrich Merz, Alexander Dobrindt und Fraktion, Antrag der Fraktion der CDU/CSU – Marktmacht von Tech-Giganten in die Schranken weisen, Deutscher Bundestag, Drucksache 20/686, 15.02.2022.
12. Alexander Mittermeier, Die 10 wichtigsten Fintech-Unternehmen der Welt, GeVestor, https://www.gevestor.de/finanzwissen/oekonomie/rankings/die-10-wichtigsten-fintech-unternehmen-der-welt.html, Stand 01.06.2022, Aufgerufen am 02.07.2023.
13. Michael E. Porter, James E. Heppelmann, Wie Smarte Produkte den Wettbewerb verändern, in Harvard Business Manager, Dezember 2014, Seite 34–60.
14. Jonas Rest, Das Ende der Illusion, manager magazin, Oktober 2022.
15. Alec Ross, Die Wirtschaftswelt der Zukunft, Plassen Verlag, 2016.
16. Philipp Alvares des Souza, Christian Schütte, La Vie en Tech, manager magazin, August 2017, Seite 72–76.

17. Christian Schütte, Die Ökonomie des Fahrdienstes Uber, manager magazin, September 2017, Seite 25.
18. Carolin Treichl, Industrie 4.0-Studie: Wie die Schweizer Wirtschaft Chancen und Risiken sieht, Presseportal, 15.08.2017, http://www.presseportal.de/pm/126139/3709131, aufgerufen am 13.09.2017.
19. Kathrin Werner, Ubers Übel, Capitel, August 2017, Seite 68–74.
20. Alibaba stellt Weichen für Expansion, Lebensmittelzeitung, Nr. 39, 29. September 2017, Seite 10.
21. Ranking der 25 wertvollsten Marken nach ihrem Marktwert im Jahr 2017, https://de.statista.com/statistik/daten/studie/6003/umfrage/die-wertvollsten-marken-weltweit, aufgerufen am 12.09.2017.
22. Studie Industrie Monitor Innovation 2014, Staufen AG, 2014.
23. Volumen der Venture Capital-Investitionen in den USA von 2009 bis 2022, www.statista.com, aufgerufen am 03.12.2023.
24. Startups: Entwicklung des Investitionsvolumens bis 2022|Statista, aufgerufen am 13.12.2023.
25. Theophane Hartmann, Frankreich: 6 Milliearden € für Technologiesektor, Frankreich: 6 Milliarden Euro für Technologiesektor – EURACTIV.de, 14.06.2023.

Quo Vadis Germany

11

Since the Hannover Messe in 2011, where the "starting signal" for Industry 4.0 was given, much has happened in Germany and its businesses. The journey towards digitalization and Industry 4.0 has begun. Although the initial steps were somewhat hesitant, the "train" has now clearly picked up speed. "The world is on the move, it does not wait for Germany," see [7]. This statement by former Federal President Roman Herzog clearly shows that things need to change in Germany. The challenges have continued to grow since 2011, and at the same time, political and economic crises have increased. After one crisis comes the next. Unfortunately, Germany is still not sufficiently prepared for the digital transformation. One example of this is public investment in its own infrastructure, especially digital infrastructure. These investments are below the European average. Countries such as France, the Netherlands, or Poland are investing significantly more money in this area, see [3]. Particularly striking is the lagging behind of the public sector, not only at the federal level but unfortunately also at the state and municipal levels.

The changes are so profound that companies cannot manage them alone. It is therefore not surprising that Bitkom President Ralf Wintergerst comes to a sobering assessment. The German economy and administration are only in the middle of the pack when it comes to digitalization, see [2].

To change this situation, the federal government supports digitalization in companies through various funding projects and through the Federal Agency for Disruptive Innovation (Sprind for short). This is intended to provide flexible and rapid state support to highly innovative start-ups that cannot be financed by investors because the respective innovation is still in its early stages. Fortunately, their budget is being continuously increased and amounted to a three-digit million sum in 2023. In addition, universities such as TU Munich are building an ecosystem for start-ups (TUM Venture Labs). The Bavarian state government supported this approach in 2022 with €2 million. Various industrial companies are also participating in this, see [15]. Such efforts by the state must

W. Huber, *Industry 4.0 Compact—How Technologies Are Transforming Our Economy and Our Companies*, https://doi.org/10.1007/978-3-658-51029-9_11

be further expanded. Regulatory obstacles must also be overcome and the reduction of bureaucracy must finally be tackled.

All sectors and thus all companies face competition from new players and, above all, new rules of the game. The belief that this "hype" will simply pass is a misconception that can and will prove fatal. A hesitant attitude on this issue can therefore have very serious consequences for the respective companies. Action is therefore required. The question is not whether a sector is affected, but rather the specific manifestation, and thus the extent and timing of the change, which varies by sector.

The pursuit of cost optimization and improved competitiveness will continue to drive digitalization forward. As a result, the importance and thus the market for data-driven services in Germany will grow in the coming years. As one of the world's leading industrial nations with a significant share of manufacturing companies, Germany must also become significantly more flexible and cost-effective in production. The current shortage of skilled workers should further fuel digital transformation. Unfortunately, this strategic component is still not sufficiently recognized in many executive suites. This is evident from the insufficient speed of change. However, it must not be overlooked that changes must be approached with care. Employees must be "brought along" (including the works council). The transformation of a company with a double-digit number of production sites across all time and cultural zones is an extremely complex and challenging undertaking. Continuous improvements must be achieved, the transfer of experience organized, and employee qualification ensured. This takes time and must also be reflected organizationally within the company. It is therefore not surprising that companies have not yet fully implemented a smart enterprise.

Customer behavior—our own included—is also being influenced by new technologies. Digitalization is changing our information and purchasing behavior, as well as the communication and interaction between customers and companies, see [10]. Companies are therefore well advised to test new customer channels. If the feedback is positive, the channel can be further expanded; if negative, the attempt can be discontinued. This means that the internal mindset within companies must also change. Experimenting with new things must be at the top of the agenda. This also entails the possibility of failure. For example, the return on investment for YouTube campaigns is higher than for TV spots. It is therefore not surprising that Google in the USA can record higher advertising revenues than all US daily newspapers combined, see [10].

11.1 Critical Review

The German economy and its defining companies are undergoing a dramatic transformation. The "first round" of digitalization has been lost. The lead of German industry over other economies has diminished, even if this is not yet directly reflected in sales figures. This new competition and change are evident, as mentioned several times, particularly in the automotive industry, Germany's flagship sector. German companies think and act too much in terms of products and production, and far too little in terms of new business models. However, it is in this area that the economic success of Germany as a business and production location will be decided. Unfortunately, the pace of transformation in the production environment is also insufficient. Thinking is often stuck in familiar and traditional patterns. The opportunities and possibilities of digitalization are still not sufficiently recognized. One reason for this is the still inadequate level of knowledge in executive suites. This is evident, for example, in the field of AI. Here, Germany boasts world-leading research. Unfortunately, there is a lack of implementation in business practice. However, this is not solely the fault of companies, but also due in part to political framework conditions. Much more intensive government support is needed here, see [1]. At the same time, management must recognize the economic potential. In many cases, these are higher than those of automation, with shorter ROIs (Return on Investment), since solutions only need to be developed once, in contrast to automation. On the other hand, automated solutions provide a good foundation for digitalization. Therefore, both approaches must be promoted, which, however, requires corresponding investments.

The realization that something must be done is becoming increasingly widespread. Large companies now cooperate as a matter of course with start-ups, or even found their own or spin off relevant ventures. For example, Bosch (a 130-year-old industrial company) is one of Germany's largest venture capital investors. The company is involved in around 30 startups worldwide and manages approximately €420 million in venture capital. Siemens is similar. Here, around €800 million in venture capital has been invested in 180 companies, see [9]. A dedicated division for this topic has even been established. Porsche, as a premium automobile manufacturer, has also joined the hunt for start-ups, establishing a hub in Israel and integrating relevant companies.

On the other hand, the strong dominance of American internet companies and their capital power should not be underestimated. In 2015, the market capitalization of US companies was $1159 billion—about 15 times larger than the entire internet economy of Germany, South Korea, and Sweden combined, and the trend is rising. The market capitalization of Alphabet alone exceeds that of all German companies in the entire digital economy. This dominance is increasing. Over the past 10 years, the market capitalization of US companies has increased about fivefold, whereas their German counterparts have only seen moderate growth, see [6]. This illustrates the dramatic nature of the situation. US companies use their available capital for further expansion. The importance of data, and thus knowledge about their customers, is much more pronounced in the US than in Germany. Moreover, these companies practice the networking of individual technologies. For example, Apple and Google entered the mobile payment systems market with Apple Pay and Google Wallet. Due to the dominance of both companies in mobile operating systems, competitors have little chance here. On the other hand, a corresponding counterweight is being built up in Europe. Furthermore, Google is expanding its activities into the energy sector. For example, in 2014, the company acquired Nest for a billion-dollar sum. In 2023, the company invested over $300 million in an AI company, see [16]. These are just a few examples that show how companies skillfully handle data and networking to further expand their own digital ecosystem. The dominance of American companies, primarily in the B2C market, is consistently being extended to the B2B market, as the examples in this book show. The various examples also demonstrate that the B2B and B2C markets and their specific customer relationships are increasingly converging. Customers now expect similar functionality and service in the business environment as they are used to in their private lives. Business processes thus follow the logic of technologies.

The Chinese market is also a prime example of a high degree of digitalization. Cash payments have now become exotic. AliPay, for example, handles all necessary activities. In general, Chinese society is much more open to innovation, especially digital innovations, than, for example, German society. German automobile manufacturers are currently having to learn this the hard way. The car is evolving into another digital asset, and seamless integration is expected. In this respect, the Chinese competitors are clearly ahead of the somewhat more sluggish German premium brands (as mentioned several times), see [5].

11.2 What Needs to Be Done?

There is certainly some positive news to report. While Germany is clearly lagging behind in the B2C sector, "Made in Germany" holds a leading position in the country's highly important B2B sector. Companies such as Bosch, SAP, and Siemens exemplify this. These companies have consistently and specifically evolved. Cloud approaches, big data, and working with start-ups are part of their strategic toolkit. For example, in 2023 and beyond, Siemens is investing over €1 billion in its German locations, with around €500 million of that going to the new campus in Erlangen. The metaverse will be a particular focus here, see [8].

Such initiatives help secure existing markets and open up new ones. Siemens is making significant efforts to support its customers on their journey of digital transformation. But there are also examples of much smaller companies, including in the B2C sector, that show Germany is making progress. Thermondo, a platform focused on heating system installation, or Tolino as an eBook reader, are just a few examples demonstrating that it is possible to successfully build platforms in Germany and compete against seemingly overwhelming rivals like Amazon. In 2015, Tolino held a market share of around 45% for eBook readers, see [10]. However, by 2022, its market share had dropped to around 30%, see [13].

There is also progress in the field of machine learning in Germany. According to a PwC study, the technology generated additional global growth of around US$16 trillion from 2013 onwards, see [11]. According to a bitkom study from 2023, the German market amounted to €6.3 billion that year. For the coming year, growth of 30% is expected, see [12].

Germany need not hide when it comes to research, as already mentioned, see [1]. The most prominent current example is generative AI, i.e., large language models. Here, McKinsey expects a global productivity increase of around US$2.6 to 4.5 trillion. Of course, this depends heavily on the specific use case and industry, see [14]. All the more encouraging is the fact that a German startup has achieved unicorn status in this field, meaning a market valuation exceeding US$1 billion. In 2022, the company Celonis even reached decacorn status, with a market valuation exceeding US$10 billion, see [1, 4]. However, this should not obscure the fact that the most important players in AI, and especially in generative AI, are based in the USA, with Microsoft, Google (Alphabet), Apple, and Amazon. Around 70% of AI models come from the USA and over 15% from China. This demonstrates the strong dominance of both countries in this field.

German car manufacturers are also taking the issue of data very seriously. This is demonstrated by the acquisition of the map maker Here (see also Chap. 10) by a consortium of German automotive manufacturers. These companies are going even further and building a corresponding digital ecosystem. BMW and Continental are stepping up their efforts. According to the plans of these two automotive heavyweights, the interface for diagnostic systems and apps will in future be managed restrictively, see [1]. In addition, efforts are being intensified to become more independent from American and Chinese companies in the area of cloud and solutions. Initially, companies were rather reluctant to get involved with Catena-X, Gaia-X, and Manufacturing-X. Currently, however, a trend reversal is emerging. BMW is taking the lead as a major car manufacturer. This is being supported by funding programs from the Federal Ministry for Economic Affairs. Taken together, these examples show that things are moving in Germany. Whether this will be enough remains to be seen.

In general, the opportunities presented by digitalization and Industry 4.0 should be seized pragmatically, rather than getting lost in academic debates. The examples show that with imagination and a strong will, it is possible to achieve results similar to those of the well-known major players from the USA or China. The positive aspect is that this applies to all companies, regardless of size. Thus, every company has its own individual opportunity, and it is important to take advantage of it. The upcoming changes should be seen as something positive.

Nevertheless, coordinated government support is absolutely essential. The current programs show that things are moving in the right direction (as already mentioned). They must also focus more than before on start-ups and medium-sized companies as sources of innovation. These are forging new paths of innovation. Start-ups in particular can become central to established companies. In connection with this, venture capital should be significantly increased. In the past, large corporations have received substantial support. This is certainly important. However, the German Mittelstand, with its strongly owner-managed companies, forms the backbone of Germany as an industrial location. In many cases, very long-term decisions and strategies are pursued here. These must be specifically supported through funding programs. The federal government must make significantly higher investments in these companies in order to remain one of the leading economies. The goal should be for Germany to become the leading nation in digitalization. This also includes, finally, a sensible and thus nationwide expansion of digital infrastructure. Furthermore, Germany should aim to take a leading role in e-government. Currently, it is the Scandinavian and Baltic states that are clearly ahead in this area. This assessment was addressed in 2017, as well as in a very similar form in 2016, by the report on "Research, Innovation and Technological Performance in Germany" by the Expert Commission on Research and Innovation (EFI), see [6, 7]. In the current report from 2022, there is a stated willingness to catch up technologically, see [6]. The currently extended or newly launched funding programs show that things are moving in the right direction. However, the intensity and scope need to be expanded.

11.3 What Happens Next

The changes currently taking place are difficult to comprehend and even harder to predict. Traditional methods of competitive analysis are hardly effective anymore or provide misleading signals. Disruptive changes and thus breaks in development are increasingly shaping our economic life, and changes are occurring at ever shorter intervals. Forecasts based on historical data are therefore increasingly producing incorrect results. As a consequence, one must not rest on the successes of the past. They often blind us to the need for a change in perspective in order to meet new challenges. For example, Amazon has dethroned the former giants IBM and Google in the field of cloud services. Currently, the former digital bookseller Amazon holds over 40% of the market share, followed by Microsoft, Google, and IBM. The established players (even those from Silicon Valley) simply did not have the competitor "on their radar," see Chap. 10. At present, Microsoft is taking the AI market by storm with its acquisition of ChatGPT. A few years earlier, the American software company attempted something similar with the topic of the Metaverse. However, the next change is already on the horizon. These changes (whether classified as evolutionary or disruptive is up to the observer/reader) will significantly accelerate in the coming years.

As the examples show, new "players and technologies" are constantly entering the field, but well-known representatives are also performing well, as the examples demonstrate. They can come from any industry and pose a lasting threat to one's own business sector. In the context of digitalization and the associated digital transformation, technological possibilities must be leveraged to further develop one's own business.

It makes no sense to try to copy America in this area, especially since Germany has no need to "hide" when it comes to production and technology development. The only area where there is a need to catch up is in digital business models. However, every industry and every company has its own unique dynamics, and the same applies to nations and economies. Companies must transform their own core competencies and strengths into the digital age (which has long since begun). This also includes employee rights. Employees are increasingly at risk of becoming interchangeable, which will increase the pressure. The goal is to shape a socially responsible transformation that does not jeopardize economic success. This also requires a changed attitude and approach to learning. The changes show that lifelong learning is becoming increasingly important for everyone (whether employees in a company, board members, or even teachers). Education should not be understood merely as the acquisition of competencies. A well-developed education system across all areas is simply a key success factor for an innovative and thus future-proof society. Therefore, existing forms of learning must be constructively questioned.

In addition, new forms of work must be promoted in companies. Since the coronavirus pandemic at the latest, working from home has become part of everyday professional life. "Workation" (working where others go on vacation) is the next evolutionary step. Companies would be well advised to further develop both in the areas of learning and qualification as well as in work design. Such approaches are simply expected by applicants.

However, digitalization is not only blurring industry boundaries, but increasingly also national borders. Therefore, a unified European single market with harmonized regulations is needed. The European General Data Protection Regulation (GDPR) points in the right direction, even if criticism persists. At the same time, one should beware of overregulation, as is currently being discussed in the field of AI. This would be counterproductive. Furthermore, a European research area with coordinated activities must be established in order to strengthen the European single market and Europe's position against challengers from the USA and Asia. In Germany, it is finally necessary to further digitalize public administration and reduce the continuously increasing regulation. This could provide important impulses for the private sector and, overall, promote the development of a "Digital Ecosystem Germany." An increase in the research and education budget is therefore essential to secure Germany as a business location. This also includes significantly increased spending on business-oriented developments and innovations, especially in new business models.

References

1. Martin Bayer, KI in Deutschland: Forschung hui – Business pfui, Computerwoche, 14.08.2023.
2. Bitkom, Deutschland bei Digitalisierung nur im Mittelfeld, CIO, 28.08.2023.
3. Max Borowski, Welche Klagen sind berechtig? Waran der Standort Deutschland wirklich leidet, ntv, 05.07.2023.
4. Yvonne Dwerne, Wefox, Trade Republic und Omio sind auf der Liste der deutschen Unicorns und nun gibt es auch das erste Decacorn, Esquire, 3. Februar 2022.
5. Christoph Hammerschmidt, Ausgesperrt, CarIT,m Sonderedition zur IAA 01 2017, Seite 32–34.
6. Dietmar Harhoff, Uschi Backes-Gellner, Uwe Cantner, Monika Schnitzer, Christoph Böhringer, Ingrid Ott, EFI Gutachten 2016 (Expertenkommission Forschung und Innovation),

Februar 2016, http://www.e-fi.de/fileadmin/Gutachten_2016/EFI_Gutachten_2016.pdf, aufgerufen am 15.09.2017. Version 2022, aufgerufen am 23.07.2023.

7. Dietmar Harhoff, Uschi Backes-Gellner, Uwe Cantner, Monika Schnitzer, Christoph Böhringer, Ingrid Ott, EFI Gutachten 2017 (Expertenkommission Forschung und Innovation), Februar 2017, http://www.e-fi.de/fileadmin/Gutachten_2017/EFI_Gutachten_2017.pdf. Aufgerufen am 21.09.2017.
8. Jürgen Hill, Siemens investiert eine Milliarde in Deutschland – Zentrum für industrielles Metaverse in Erlangen geplant, Computerwoche 17.07.2023.
9. Kurt Matzler, Frank Bailom, Stefphan Friedrich von den Eichen, Markus Anschober, Digitale Disruption: Wie Sie Ihr Unternehmen auf das Digitale Zeitalter vorbereiten, Verlag Fran Vahlen, 2016.
10. Jürgen Meffert, Heribert Meffert, Eins oder null, Econ Verlag, April 2017.
11. Philipp Alvares de Souza Soares, Michael O.R. Kröher, Chiristian Schütte, Brain Valley, manager magazin, August 2017, Seite 78–79.
12. Andreas Streim, Florian Bayer, Kai Pascal Beerlink, Presseinformation: Deutscher KI-Markt wächst um ein Drittel, 29.11.2023.
13. Was ist besser? Kindle oder Tolino, https://www.was-ist-besser.net/, 06. Juni 2022, aufgerufen am 22.07.2023.
14. McKinsey&Company, Press Release, Studie: Generative KI kann zum Produktivitätsbooster werden, 13. Juli 2023.
15. TUM Campus, Bayern fördert TUM Venture Labs, 09.05.2022.
16. Daniel Ziegener, Google investiert mindestens 300 Millionen in KI-Unternehmen, golem.de, 06.02.2023.

MIX
Papier aus verantwortungsvollen Quellen
Paper from responsible sources
FSC® C105338

If you have any concerns about our products,
you can contact us on
ProductSafety@springernature.com

In case Publisher is established outside the EU,
the EU authorized representative is:
Springer Nature Customer Service Center GmbH
Europaplatz 3, 69115 Heidelberg, Germany

Printed by Libri Plureos GmbH
in Hamburg, Germany